Volume 63

Ion Channel Diseases

Advances in Genetics, Volume 63

Serial Editors

Jeffrey C. Hall
Orono, Maine

Jay C. Dunlap
Hanover, New Hampshire

Theodore Friedmann
La Jolla, California

Veronica van Heyningen
Edinburgh, United Kingdom

Volume 63

Ion
Channel
Diseases

Edited by

Guy Rouleau

CHU Sainte-Justine Research Centre
University of Montreal
Quebec, Canada

Claudia Gaspar

Centre of Excellence in Neuromics
CHUM Research Centre
Notre-Dame Hospital Montreal
Quebec, Canada

ELSEVIER

AMSTERDAM • BOSTON • HEIDELBERG • LONDON
NEW YORK • OXFORD • PARIS • SAN DIEGO
SAN FRANCISCO • SINGAPORE • SYDNEY • TOKYO
Academic Press is an imprint of Elsevier

Academic Press is an imprint of Elsevier

525 B Street, Suite 1900, San Diego, CA 92101-4495, USA
30 Corporate Drive, Suite 400, Burlington, MA 01803, USA
32 Jamestown Road, London, NW1 7BY, UK
Linacre House, Jordan Hill, Oxford OX2 8DP, UK
Radarweg 29, PO Box 211, 1000 AE Amsterdam, The Netherlands

First edition 2008

ISBN: 978-0-12-374527-9
ISSN: 0065-2660

For information on all Academic Press publications
visit our website at elsevierdirect.com

Printed and bound in USA

08 09 10 11 12   10 9 8 7 6 5 4 3 2 1

Contents

Contributors   ix

Part I: Muscle Channelopathies   1

1 **Periodic Paralysis   3**
   Bertrand Fontaine
      I. Introduction   4
      II. Hypokalemic Periodic Paralysis   4
      III. Hyperkalemic Periodic Paralysis   11
      IV. Other Types of Periodic Paralysis   14
      V. Functional Studies of Muscle Ion Channel
         Mutations in Patients   15
      VI. Treatment and Care of hypoPP Patients   17
      VII. Conclusion   18
         References   18

2 **Myotonia Congenita   25**
   Christoph Lossin and Alfred L. George, Jr.
      I. Introduction   26
      II. Clinical Aspects of Myotonia Congenita   26
      III. Molecular Genetics of Myotonia Congenita   32

IV. Physiological Basis of Myotonia Congenita   40
V. Myotonia Congenita in Animal Models   45
VI. Treatment of Myotonia Congenita   46
VII. Concluding Remarks   48
References   49

3 **Familial Hemiplegic Migraine**   57
Curtis F. Barrett, Arn M. J. M. van den Maagdenberg,
Rune R. Frants, and Michel D. Ferrari

I. Introduction   58
II. The Migraine Attack: Clinical Phases and
Pathophysiology   61
III. The Migraine Aura and Cortical Spreading Depression   61
IV. Migraine as a Genetic Disorder   63
V. Familial Hemiplegic Migraine: A Model for
Common Migraine   65
VI. Functional Consequences of FHM Mutations   68
VII. FHM as an Ionopathy: Identifying a Common
Theme Among FHM Subtypes   73
VIII. Concluding Remarks   75
References   75

4 **Genetics and Molecular Pathophysiology of
$Na_v1.7$-Related Pain Syndromes**   85
Sulayman D. Dib-Hajj, Yong Yang, and
Stephen G. Waxman

I. Introduction   86
II. Role of $Na_v1.7$ in Pain Syndromes: Animal Studies   90
III. $Na_v1.7$ and Inherited Pain Syndromes   92
IV. Conclusions   104
References   104

Part II: Internal Diseases   111

5   **Channelopathies of Transepithelial Transport and Vesicular Function**   113
   Christian A. Hübner and Thomas J. Jentsch
   I. Introduction   114
   II. Disorders   116
   III. Concluding Remarks   142
   References   142

   Index   153

CONTRIBUTORS

Numbers in parentheses indicate the pages on which the authors' contributions begin.

Curtis F. Barrett (57) Department of Neurology, Leiden University Medical Center, Leiden, The Netherlands; Department of Human Genetics, Leiden University Medical Center, Leiden, The Netherlands

Sulayman D. Dib-Hajj (85) Department of Neurology, Yale University School of Medicine, New Haven, Connecticut 06510; Center for Neuroscience and Regeneration Research, Yale University School of Medicine, New Haven, Connecticut 06510; Rehabilitation Research Center, VA Connecticut Healthcare System, West Haven, Connecticut 06516

Michel D. Ferrari (57) Department of Neurolgy, Leiden University Medical Center, Leiden, The Netherlands

Bertrand Fontaine (3) INSERM, UMR 546, Paris, France; Université Pierre et Marie Curie-Paris 6, UMR S546 and Assistance Publique-Hôpitaux de Paris, Centre de référence des canalopathies musculaires, Groupe Hospitalier Pitié-Salpêtrière, Paris, France

Rune R. Frants (57) Department of Human Genetics, Leiden University Medical Center, Leiden, The Netherlands

Alfred L. George Jr. (25) Departments of Medicine and Pharmacology, Vanderbilt University School of Medicine, Nashville, Tennessee 37232

Christian A. Hübner (113) Department of Clinical Chemistry, University Hospital of the Friedrich-Schiller-Universität, Erlanger Allee 101, D-07747 Jena, Germany

Thomas J. Jentsch (113) FMP (Leibniz-Institut für Molekulare Pharmakologie) and MDC (Max-Delbrück-Centrum für Molekulare Medizin), Robert-Rössle-Strasse 10, D-13125 Berlin, Germany

Christoph Lossin (25) Department of Neurology, UC Davis School of Medicine, Sacramento, California 95817

Arn M. J. M. van den Maagdenberg (57) Department of Neurology, Leiden University Medical Center, Leiden, Netherlands; Department of Human Genetics, Leiden University Medical Center, Leiden, The Netherlands

Yong Yang (85) Department of Dermatology, Peking University First Hospital, Beijing 100034, China

Stephen G. Waxman (85) Department of Neurology, Yale University School of Medicine, New Haven, Connecticut 06510; Center for Neuroscience and Regeneration Research, Yale University School of Medicine, New Haven, Connecticut 06510; Rehabilitation Research Center, VA Connecticut Healthcare System, West Haven, Connecticut 06516

Part I

MUSCLE CHANNELOPATHIES

1

Periodic Paralysis

Bertrand Fontaine

INSERM, UMR 546, Paris, France; Université Pierre et Marie Curie-Paris 6, UMR S546 and Assistance Publique-Hôpitaux de Paris, Centre de référence des canalopathies musculaires, Groupe Hospitalier Pitié-Salpêtrière, Paris, France

I. Introduction
II. Hypokalemic Periodic Paralysis
III. Hyperkalemic Periodic Paralysis
IV. Other Types of Periodic Paralysis
V. Functional Studies of Muscle Ion Channel Mutations in Patients
VI. Treatment and Care of hypoPP Patients
VII. Conclusion
Acknowledgments
References

ABSTRACT

Periodic paralyses are rare diseases characterized by severe episodes of muscle weakness concomitant to variations in blood potassium levels. It is thus usual to differentiate hypokalemic, normokalemic, and hyperkalemic periodic paralysis. Except for thyrotoxic hypokalemic periodic paralysis and periodic paralyses secondary to permanent changes of blood potassium levels, all of these diseases are of genetic origin, transmitted with an autosomal-dominant mode of inheritance. Periodic paralyses are channelopathies, that is, diseases caused by mutations in genes encoding ion channels. The culprit genes encode for potassium, calcium, and sodium channels. Mutations of the potassium and calcium channel genes cause periodic paralysis of the same type (Andersen–Tawil syndrome or

Advances in Genetics, Vol. 63
0065-2660/08 $35.00
DOI: 10.1016/S0065-2660(08)01001-8

hypokalemic periodic paralysis). In contrast, distinct mutations in the muscle sodium channel gene are responsible for all different types of periodic paralyses (hyper-, normo-, and hypokalemic). The physiological consequences of the mutations have been studied by patch-clamp techniques and electromyography (EMG). Globally speaking, ion channel mutations modify the cycle of muscle membrane excitability which results in a loss of function (paralysis). Clinical physiological studies using EMG have shown a good correlation between symptoms and EMG parameters, enabling the description of patterns that greatly enhance molecular diagnosis accuracy. The understanding of the genetics and pathophysiology of periodic paralysis has contributed to refine and rationalize therapeutic intervention and will be without doubts the basis of further advances. © 2008, Elsevier Inc.

I. INTRODUCTION

The recognition of periodic paralysis (PP) as a distinct entity and the proposal of its muscle origin are contributions of the German medical school of the late nineteenth century. In the first middle of the twentieth century, physicians from Europe and America further observed that modifications of the blood potassium levels occurred during attacks (historical review in Buruma and Schipperheyn, 1979), leading to a usable definition of the disease in clinical practice. Nowadays, PP is still classified according to the concomitant blood potassium variations into hypokalemic (hypoPP), normokalemic (normoPP), and hyperkalemic periodic paralysis (hyperPP). What has changed in the last 20years is the recognition of these diseases as channelopathies and the understanding, which is far from complete, of their mechanisms, giving a rationale to therapeutic intervention.

II. HYPOKALEMIC PERIODIC PARALYSIS

HypoPP is characterized by reversible attacks of muscle weakness associated with decreased blood potassium levels (Fontaine et al., 2007; Venance et al., 2006). There are several causes for this disorder. HypoPP can be secondary to renal or gastrointestinal potassium loss. In this occurrence, abnormal blood potassium levels are also observed between attacks, and muscle weakness fluctuates in parallel with blood potassium levels. Muscle weakness is indeed thought to be directly related to the degree of muscle depolarization induced by hypokalemia. In primary hypoPP, blood potassium levels are abnormal only during attacks. Two forms of primary hypoPP have been recognized: (1) thyrotoxic hypoPP which is associated with thyrotoxicosis and (2) familial hypoPP which is a genetic disorder of autosomal-dominant inheritance.

Thyrotoxic hypoPP is considered to be secondary to hyperthyroidism (Kung, 2006). The frequency of thyrotoxic hypoPP is not known (only a small percentage of patients with thyrotoxicosis will develop hypoPP). It is present in all populations but occurs more frequently among Asians. Most cases are sporadic with a male predominance (9:1), but familial cases have been anecdotally described. Patients usually present with paralysis of the four limbs and a profound decrease in blood potassium levels (down to less than 1 mEq/l). Attacks may spontaneously recover and recur if thyrotoxicosis is not treated. In a number of cases, signs of hyperthyroidism are clinically obvious, but this is not always the case. In Caucasians, signs and symptoms of thyrotoxicosis are often absent. Attacks of hypoPP occur only during the state of hyperthyroidism and never when the thyroid function is back to normal. Therefore, the treatment is based upon the restoration of the euthyroid state. In the meantime, potassium chloride may be given to improve muscle force and prevent the occurrence of new attacks. Some patients may also have some degree of glucose intolerance.

Thyrotoxic hypoPP is thought to be related to an increased activity of the sodium potassium ATPase (Na/K-ATPase) pump, which presumably causes an intracellular shift of potassium and hypokalemia (Kung, 2006). A tentative hypothesis is that thyrotoxic hypoPP patients have a predisposition to the Na/K-ATPase pump activation by thyroid hormones or hyperinsulinism.

The alternate form of primary hypoPP is familial hypoPP, commonly referred to as hypoPP. The exact frequency of the disease is unknown: the prevalence has been estimated at 1:100,000 in Danish registries (Buruma and Schipperheyn, 1979).

Familial hypoPP is a monogenic disorder with an autosomal-dominant mode of inheritance. Some cases may present as sporadic because of the incomplete penetrance of the disease, mostly in women (Elbaz *et al.*, 1995; Links, 1992). The disease is potentially lethal. A mortality of 10% was common before the introduction of intensive care units. Nevertheless, even today, severe cases may still occur and patients with hypoPP should be cared with caution in intensive care units mostly when blood potassium levels are low (Caciotti *et al.*, 2003).

Although the genetic defect is present throughout life, the mean age at onset is in the second decade. The frequency of attacks is higher from the second to the fourth decades of life, and then tends to decrease. Attack frequency is very variable, ranging from once in a lifetime to several per week. Affected women tend to have fewer attacks than affected men, reflecting the decreased expressivity of the disease in women.

Attacks of muscle weakness usually affect the four limbs. When incomplete, they often predominate in the lower limbs. Respiration, deglutition, and ocular motility are usually spared, but may be affected in the most severe attacks. Coughing is more difficult during attacks. Attacks usually occur several hours after strenuous exercise or a meal rich in carbohydrates. Typically, patients wake in the night paralyzed. Attacks may also be milder affecting one or more limbs. Attacks last several hours and resolve spontaneously. Recovery may be sped up by the ingestion of potassium chloride and may be aborted by pursuing exercise at a moderate level. If blood potassium levels are measured during an attack, they are found to be below the normal range. For diagnostic purposes, measurements of blood potassium levels are important. They can be performed in the emergency room or at home. Since serum potassium levels are normal between attacks, it is important for confirming the diagnosis to obtain measurements during an attack.

It is now well established that hypoPP does not affect the heart per se. Cardiac arrhythmias may be provoked by severe hypokalemia but the heart muscle as well as smooth muscles are not affected by the disease, as shown in two autopsied cases (Links, 1992). It can be hypothesized that reported cases of hypoPP with persistent electrocardiogram abnormalities between attacks were actually Andersen–Tawil syndrome (Tawil et al., 1994).

A number of hypoPP patients develop permanent muscle weakness. The frequency of patients with permanent muscle weakness between attacks is not known. The presence of permanent weakness is well known to clinicians, as well as the variability of weakness from one examination to another. The significance of this permanent muscle weakness has been and is still debated as discussed below (Links et al., 1990).

Permanent muscle weakness is rarely but possibly observed in young adult age. It usually fluctuates. The patients are generally aware that they can partially control the intensity of the weakness by mild exercise, or by the ingestion of potassium chloride salts or acetazolamide tablets. In this circumstance, permanent weakness may indicate a state of persistent or continuously recurring mild attacks and may be dramatically improved by long-term use of carbonic anhydrase inhibitors (Buruma and Schipperheyn, 1979; Dalakas and Engel, 1983; Griggs et al., 1970; Links et al., 1988).

In contrast, persistent muscle weakness is less fluctuating in older patients and is less sensitive to medication. This lack of responsiveness to treatment is a sign of muscle degeneration, a fixed myopathy. The frequency of the myopathy is not well established. This is due to the fact that hypoPP patients rarely undergo a muscle biopsy and there is no reliable clinical imaging method to detect such a myopathy (Links, 1992). Some authors have proposed that it affects all patients over 50years old to some degree (Links et al., 1990). Its severity is variable, ranging from weakness evidenced only at clinical examination but not interfering with a normal life in most patients, to wheelchair bound patients. The myopathy arises independently of the frequency and the severity of the attacks. It can even occur in the absence of attacks (Buruma and Bots, 1978; Links et al., 1990). The onset is usually in the fourth or fifth decade of life. The myopathy affects mostly the muscles of the pelvic girdle and the proximal muscles of upper and lower limbs.

Vacuoles are regularly seen on muscle biopsies and are considered the hallmark of the myopathy, which is termed a *vacuolar myopathy*. They are thought to be markers of the myopathy, although vacuoles are seen in patients without a permanent muscle weakness. Vacuolar myopathy affects mostly muscles of the pelvic girdle and proximal muscles of upper and lower limbs. The origin of the vacuoles has been debated. Pathological studies suggest that the vacuoles arise from the sarcoplasmic reticulum and the tubular system as the result of proliferation and degeneration of these membranous organelles (Engel, 1966, 1970). Several stages have been distinguished by pathological studies with different degrees of muscle fiber degeneration. Other morphological changes are also observed in periodic paralysis. Tubular aggregates which originate for the endoplasmic reticulum are localized under the muscle fiber membrane. Other changes include an increase in central nuclei, abnormal variation in fiber size, fiber necrosis, and proliferation of connective tissue. There is no animal model of vacuolar myopathy. Therefore, the exact sequence of events and the pathophysiological link between the formation of vacuole and muscle degeneration are still not understood. HypoPP represents thus a good model to understand the relationship between an ion channel mutation, abnormal cellular excitability, and cell death.

Since there was no correlation in terms of severity between attacks and myopathy, it was proposed that attacks (periodic paralysis) and myopathy (evidenced by a fixed permanent weakness) may represent two independent phenotypes with different ages at onset, and varying penetrance and expressivity (Elbaz *et al.*, 1995; Links *et al.*, 1994a). The factors, genetic or environmental, triggering and controlling the expression of these phenotypes are unknown.

Several causes have been advocated for hypoPP, such as a defect in potassium metabolism. In the late 1980s, the group of Rüdel and Lehmann-Horn in Germany developed a technique to record the electrophysiological activity from muscle biopsies of patients with neuromuscular diseases (Rudel *et al.*, 1984). This technique enabled them to make the seminal observation that when decreasing the potassium concentration in the extracellular medium, they could depolarize the muscle cells, which correlated with paralysis (Rudel *et al.*, 1984). In contrast, normal muscle cells became hyperpolarized when the extracellular potassium was lowered. This observation suggested that abnormal ion fluxes might be implicated, a hypothesis which was reinforced when the primary role of a sodium channel was demonstrated in the other form of periodic paralysis: hyperPP (Fontaine *et al.*, 1990; Ptacek *et al.*, 1991; Rojas *et al.*, 1991). The same strategy used to identify the sodium channel in hyperPP was applied to hypoPP. Large families were collected to enable linkage analysis. The approach was based on the fact that each gene is present in the genome at two copies with different forms (alleles). The segregation of the alleles of the gene can be traced through generations. The cosegregation of one allele of a specific gene with the disease indicates that the tested gene is causative (linkage). In the case of hypoPP, the most obvious candidates were potassium channels. It became quickly obvious that none of the alleles of potassium channels segregated with the disorder. The study was then enlarged with markers covering the whole genome. Surprisingly, the first genetic markers which cosegregated with the disease were in close vicinity with a voltage-gated calcium channel (Fontaine *et al.*, 1994). Soon after, mutations were found in the voltage-gated calcium channel *CACNA1S*, establishing it as the first hypoPP causing gene (Fig. 1.1) (Jurkat-Rott *et al.*, 1994; Ptacek *et al.*, 1994). The voltage-gated calcium channel is made of four homologous domains, each of them composed of six transmembrane segments. Remarkably, all mutations changed positively charged amino acids arginines in the voltage sensor segment 4 (R528H or G in domain II, as well as R1239H of G in domain IV). The phenotype associated with both mutations is grossly similar (Elbaz *et al.*, 1995; Fouad *et al.*, 1997). The penetrance is incomplete and tends to be lower in women and to depend on ethnicity, although this latter point needs confirmation (Elbaz *et al.*, 1995; Fouad *et al.*, 1997; Ikeda *et al.*, 1996; Kawamura *et al.*, 2004; Miller *et al.*, 2004; Sillen *et al.*, 1997). These two mutation sites in the calcium channel which cause hypoPP have been now found in all studied populations (review in Fontaine *et al.*, 2007). Mutations have only been observed in familial hypoPP but not in thyrotoxic hypoPP, although there are anecdotal reports of familial hypoPP revealed by thyrotoxicosis (review in Fontaine *et al.*, 2007).

Electrophysiological patterns of most frequent responses to exercise tests and EMG recordings			
Clinical phenotype		HyperPP	HypoPP
Channel mutations		T704M sodium	R528H calcium
Electrophysiological pattern		IV	V
Needle-EMG	Myotonic discharges	No or rare	No
CMAPs after short exercise	PEMP	No	No
	Amplitude change after first trial	Increase	No
	Amplitude change after second or third trial	Gradual increase	No
CMAPs after long exercise	Immediate change of amplitude	Increase	No
	Late change of amplitude	Decrease	Decrease
Sensitivity[a] (%)		83	84

Figure 1.1. Classification of periodic paralyses according to electromyography.

The voltage-gated calcium channel lies within the T-tubules, which are intracellular invaginations of the muscle membrane. Their known role is to couple the action potential with the intracellular release of calcium from the sarcoplasmic reticulum through the activation of calcium channels termed *ryanodine receptors*. In hypoPP muscle fiber, the excitation–contraction appears to be normal (Engel and Lambert, 1969; Ruff, 1991). Expression studies of mutated calcium channels have been technically difficult because of the poor expression of the muscle calcium channel in *in vitro* systems. They have shown minor abnormalities pointing to a loss-of-function effect (decreased current density and slowed activation) (Lapie *et al.*, 1996; Morrill and Cannon, 1999). Other studies on muscle fibers obtained from biopsies have pointed to a reduction of an inwardly rectifying or an ATP-sensitive potassium current (Ruff, 1999; Tricarico *et al.*, 1999). Thus, although the mutations in the calcium channel have been known for more than 10 years, there is no clear understanding of how they could provoke muscle fiber paralysis induced by depolarization and hypokalemia (Cannon, 2006).

Genetic linkage studies rapidly established that hypoPP was genetically a heterogeneous disease. The study of large families, demonstrated by linkage analysis and mutation search, showed that mutations in the voltage-gated sodium channel also caused hypoPP (Bulman *et al.*, 1999; Jurkat-Rott *et al.*, 2000). This discovery came as a surprise since it was already known that mutations in the voltage-gated sodium channel caused other forms of periodic paralysis (Fig. 1.1).

The voltage-gated sodium channel belongs to the same channel family as the calcium channel and shares a similar organization. It is made of four homologous domains (domains I–IV), each of them composed of six transmembrane segments (S1–S6). It is the key player in the action potential being responsible for the firing and propagation of action potentials. It is opened in a voltage-gated manner (activation) by membrane depolarization and subsequently closed (fast inactivation), thus terminating action potentials with an additional contribution of potassium currents generated by voltage-gated potassium channels. Segments S4 are made up of a chain of one positive amino acids (mostly arginines) and two hydrophobic residues which anchor the segment in the membrane. By being attracted toward the extracellular side of the membrane, the positive amino acids open the pore of the channel. The intracellular loop between domains III and IV is responsible for the (fast) inactivation of the channel. This loop is attracted by the sodium flux and mechanically closes the pore by interacting with amino acids of the channel positioned on the intracellular side of the membrane. Mutations causing hypoPP notably lie in different regions of the sodium channel than those causing other forms of periodic paralysis (Fig. 1.1). Interestingly, they affect similar amino acids to those mutated in the calcium channel. They indeed change arginines in position 669 and

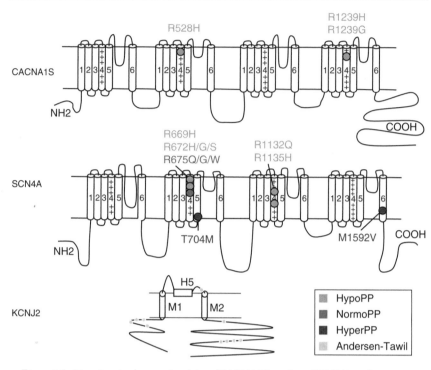

Figure 1.2. Mutations in the muscle calcium (*CACNA1S*), sodium (*SCN4A*), and potassium (*KCNJ2*) genes causing periodic paralyses.

672 in the voltage sensor S4 of domain II, as well as in position 1132 in the voltage sensor S4 of domain III (Carle *et al.*, 2006). These mutations have been found in all populations studied so far but at a lower frequency than the one observed for calcium channel mutations (10% vs 60%) (review in Fontaine *et al.*, 2007). The penetrance of sodium channel mutations is probably higher than the one observed for calcium channel mutations, at least in the only large family studied so far (Sternberg *et al.*, 2001). Differences have been noted in the phenotype displayed by patients bearing calcium or sodium channel mutations. In patients with sodium channel mutations, hypoPP tends to begin later, is accompanied by muscle aches, shows a predominance of tubular aggregates compared with vacuoles in the muscle biopsy, and is aggravated by acetazolamide (Sternberg *et al.*, 2001).

The mutations in the sodium channel causing hypoPP have been introduced by *in vitro* mutagenesis in the channel, and both mutant and control channel have been expressed in *in vitro* systems. Biophysical parameters have shown different abnormalities, an enhanced inactivation being most frequently observed (Fig. 1.2) (review in Fontaine *et al.*, 2007). Accordingly, recording of muscle fibers obtained from muscle biopsies showed a reduced current density and a slower upstroke and decreased action potentials when compared with controls. The reduction in current density is at least partly explained by abnormal channel inactivation and causes a loss of function of the membrane. Inactivation is a unique property of sodium channels which contributes to end and regulate action potentials: sodium channel close after opening regardless of membrane potential. Fast inactivation occurs in the range of milliseconds and directly contributes to terminate action potentials, whereas slow inactivation is active within seconds and regulates the number of channel available for activation. HypoPP mutations in the sodium channel were shown to enhance fast or slow inactivation, thus leading to paralysis by reducing the number of channel available for depolarization (Fig. 1.2) (Ruff, 2008). As for the calcium channel, the understanding of the mechanisms of muscle fiber paralysis and hypokalemia is far from complete.

III. HYPERKALEMIC PERIODIC PARALYSIS

HyperPP is defined by the occurrence of episodic attacks of generalized weakness (sometimes focal), accompanied by an increase in serum potassium blood levels. The disease is only truly paralytic in familial hyperPP, commonly referred to as hyperPP. A certain degree of muscle weakness can indeed be observed in disorders causing a permanent state of hyperkalemia (secondary hyperPP caused by renal or endocrine diseases), but its intensity is usually less than the one observed in familial hyperPP. Differentiating secondary and familial hyperPP is usually easy since the increase in blood potassium levels only takes place during attacks in familial hyperPP. Moreover, in secondary hyperPP, muscle weakness is less variable (less episodic) than in primary hyperPP. In secondary hyperPP, muscle paralysis is thought to be directly related to the state of membrane depolarization induced by hyperkalemia.

The prevalence of primary hyperPP is unknown but is considered to be similar to hypoPP. In familial hyperPP, hyperkalemia is far from constant since it may lack in 50% of patients as shown by studies in well-characterized hyperPP families (Feero et al., 1993; Lehmann-Horn et al., 1993; Plassart et al., 1994; Ptacek et al., 1993). In these cases, the serum potassium was or had variations within the normal range. A more consistent finding which relates potassium blood levels to paralysis is weakness provocation by ingestion of potassium salts (potassium challenge). This test was formerly used for diagnosis before it was replaced by a combination of exercise electromyography (EMG) and molecular biology (Fig. 1.3). The fact that potassium ingestion provokes attacks is useful information in terms of pathophysiology since it clearly demonstrates at bedside a link between muscle membrane depolarization and paralysis.

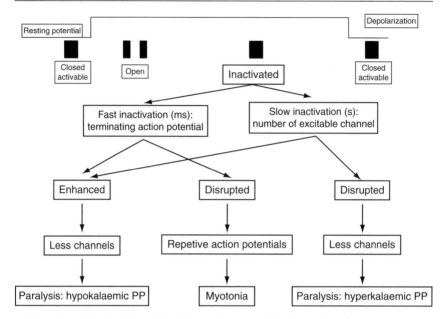

Figure 1.3. Pathophysiology of periodic paralyses caused by sodium channel gene mutations.

Observation of pedigrees showed that an almost complete penetrance is usually the rule in hyperPP but there are variations in the severity of the disease in individuals both between and within families. The disease tends to be milder in women than in men (decreased expressivity in women) (Buruma and Schipperheyn, 1979).

Attacks begin during the first decade of life. The episodes vary in frequency from several in a day to once a while. They last minutes to several hours and then remit spontaneously. In comparison with hypoPP, hyperPP tends to have an earlier onset, more frequent attacks, but much shorter and milder. Attacks are environmentally triggered. The most frequent triggering factors are rest after exercise, fasting and cold exposure. Other provocation factors depend on individuals and may include ethanol ingestion, stress, ingestion of food with high-K content, and pregnancy. Patients quickly learn that keeping exercising after the onset of an attack alleviates muscle weakness or even aborts it. At difference with hypoPP, myotonia may be associated with paralytic attacks. Myotonia is a prolonged failure of muscle decontraction. When present, it is a permanent manifestation (and not episodic) which is usually mild (compared with truly myotonic syndromes). The patient may report cramps or muscle stiffness during exercise or when exposed to cold. These symptoms can rarely been painful. A frequent sign revealed at neurological examination is myotonia of the eyelids which can be evidenced by asking the patient to close and open repeatedly his eyes. A lid-lad sign is also common: when asked to look downward, the sclera of the eyes of the patient is visible because of eyelid myotonia. In the most frequent types of hyperPP, myotonia is clinically evidenced in 1/8 of patients and by EMG in 1/2 (Plassart et al., 1994). Similar both clinically and pathologically to hypoPP, a myopathy may develop with an onset in the fourth or fifth decade of life (Bradley et al., 1990; Feero et al., 1993; Plassart et al., 1994). The myopathy may cause a permanent muscle weakness. As in hypoPP, there is no relationship between the number and the frequency of the attacks and the myopathy. Muscle weakness predominates in the proximal segment of the lower limbs and may be severe enough to necessitate assistance for walking or even a wheelchair (Bradley et al., 1990). Similar to hypoPP, muscle biopsy shows nonspecific destruction of the muscle fibers associated with vacuoles, which originate from T-tubules and endoplasmic reticular system.

From an historical point of view, hyperPP has a particular place because it is the first human disorder shown to be caused by mutations in a voltage-gated ion channel. These discoveries led to the definition of a new type of diseases caused by ion dysfunction, the so-called channelopathies. The role of the voltage-gated sodium channels in hyperPP was suggested by the observation of a noninactivating sodium current in muscle fibers from hyperPP patients, when the extracellular potassium concentrations were raised above normal values (Fontaine et al., 1990). This observation led to the hypothesis that the sodium channel might be the site of the primary defect in hyperPP. The answer to this question came in two steps: (1) linkage studies established a link between the voltage-gated sodium channel gene, SCN4A, and hyperPP (Fontaine et al., 1990) and (2) mutations were found in the coding sequence of SCN4A gene establishing it definitively as the hyperPP gene (Ptacek et al., 1991; Rojas et al., 1991). Two of the mutations causing hyperPP (Fig. 1.1), Thr704Met, and Met1592Val account for the majority of patients (Feero et al., 1993; Lehmann-Horn et al., 1993; Plassart et al., 1994; Ptacek et al., 1993).

To gain insight into disease pathophysiology, sodium channel mutations causing hyperPP were expressed into an in vitro system and studied by patch-clamp analysis. The Thr704Met and the Met1592Val mutations were introduced into cDNAs expressed in a human embryonic kidney cell line (Cannon and Strittmatter, 1993; Cummins and Sigworth, 1996; Hayward et al., 1999). Expression studies showed that both mutations disrupted inactivation without affecting single channel conductance. The Thr704Met mutation displayed late first openings, prolonged open times, and shifted the voltage dependence of activation toward negative potentials (Cannon, 2006; Cannon et al., 1993). This mutation also modified the slow inactivation. Disruption of slow inactivation is probably one of the mechanisms that cause paralysis because it is found in a majority if not all sodium channel mutations that cause PP

(Fig. 1.3) (Ruff, 2008). According to this model, muscle paralysis would be caused by a decreased number of available sodium channels for firing action potentials. There might be, however, other mechanisms as shown by theoretical reconstructions (Cannon *et al.*, 1993) or even exceptions to this rule as claimed by some authors (Jurkat-Rott and Lehmann-Horn, 2007).

IV. OTHER TYPES OF PERIODIC PARALYSIS

The existence of normoPP has been questioned because approximately 50% of patients carrying a sodium channel known to cause hyperPP such as the T704M and M1592V present measurable blood potassium levels (Chinnery *et al.*, 2002; Lehmann-Horn *et al.*, 1993; Plassart *et al.*, 1994). The study of large cohorts of patients with PP has, however, permitted the description of unique forms of PP which are truly normokalemic such as PP caused by mutations at codon 675 of the muscle sodium gene *SCN4A* (Fig. 1.1) (Vicart *et al.*, 2004). This PP is also unusual in the fact that patients can present attacks of hypoPP provoked by the ingestion of corticosteroids or by thyrotoxicosis (Vicart *et al.*, 2004). These hypoPP attacks are rare and seem to be only related to these triggering factors. The episodes of normoPP are more frequent and induced by more classical triggers such as exercise.

Since then, rare families with mixed phenotypes and voltage-gated muscle sodium channel mutations have been reported, showing that the border between hypoPP, hyperPP, and sodium channel-related myotonia periodic paralysis is not as tight as previously thought (Brancati *et al.*, 2003; Davies *et al.*, 2002; Kelly *et al.*, 1997; Kim *et al.*, 2001; Okuda *et al.*, 2001; Plassart *et al.*, 1996; Ricker *et al.*, 1986; Sugiura *et al.*, 2000; Wagner *et al.*, 1997).

It is now well established that mutations of potassium channel genes are responsible for periodic paralysis. Mutations in the potassium channel gene *KCNJ2* have indeed been established as one of the causes of Andersen–Tawil syndrome, which is associated with distinctive facial features, and cardiac conduction abnormalities (Plaster *et al.*, 2001; Tawil *et al.*, 1994).

The implication of potassium channels causing other forms of periodic paralysis has been debated. The *KCNE* genes represent a family of genes encoding for Mink-related peptides which are single transmembrane proteins, inactive on their own, but which bind and modulate biophysical properties of potassium channel pore-forming units. Based on a preferential expression in skeletal muscle, Abbott and collaborators proposed that Mink-related peptide 2 (from *KCNE3*) forms a complex with Kn3.4, a *Shaw*-type potassium channel, which results in a complex that recapitulated biophysical and pharmacological properties of potassium currents implicated in the resting potential in skeletal muscle. Abbott and collaborators also considered *KCNE3* as a candidate for hypoPP

(Abbott *et al.*, 2001). The authors looked for DNA variants in the *KCNE3* gene and found a missense mutation which changed an arginine into a histidine at amino acid 83. This missense mutation was absent from 120 control individuals. It was present in three members of a family with hypoPP and segregated with the disease within the family. The same mutation was also present in two members of another family with an unspecified type of periodic paralysis. Moreover, when expressed in rodent skeletal muscle cell, the mutant changed the excitability of muscle cells by producing depolarization. In support to this first observation, the mutation was also found in 1/15 patients with the thyrotoxic form of hypoPP (Dias Da Silva *et al.*, 2002). However, the latter observation was not reproduced in the Chinese population (Tang *et al.*, 2004).

At first glance, these observations seem to be solid enough to establish *KCNE3* as a causative gene for hypoPP. However, it must be noted that they were established in too small a number of patients to enable statistical analysis and that the number of controls was low, even if considered as adequate under current standards. The study of large collections of controls and patients changed the view on the role of *KCNE3* in hypoPP. Indeed, no other variant of the *KCNE3* gene was ever associated with periodic paralysis in contrast to other ion channel genes. In addition, the mutation was found to be associated in the same patient with a sodium channel mutation well known to cause hypoPP (Sternberg *et al.*, 2003). The phenotype of the patient who had both a potassium channel and a sodium channel mutation was similar to that of his father who carried only the sodium channel mutation, demonstrating that the potassium channel mutation did not breed true in contrast to the sodium channel mutation. The R83H mutation was also found at the same frequency in large groups of controls or periodic paralysis patients (approximately 1%) (Jurkat-Rott and Lehmann-Horn, 2004; Sternberg *et al.*, 2003). Altogether, these observations argue against a causative role of the R83H mutation of the *KCNE3* gene in periodic paralysis. The R83H mutation should therefore be considered as a variant of unknown role.

V. FUNCTIONAL STUDIES OF MUSCLE ION CHANNEL MUTATIONS IN PATIENTS

The functional consequences of ion channel mutations on muscle membrane excitability in patients can be studied by the noninvasive technique of EMG. During attacks of PP, the muscle membrane has been shown to be depolarized and unable to respond to electrical stimulation (Buruma and Schipperheyn, 1979; Links and van der Hoeven, 2000; Links *et al.*, 1994b; Zwarts *et al.*, 1988). Between attacks, the muscle membrane activity recorded by EMG is normal, albeit muscle conduction velocities are decreased. Muscle conduction velocities have been recorded by invasive and noninvasive techniques. Invasive techniques have been shown to be more sensitive but less feasible in a daily practice. Interestingly, muscle conduction velocities were shown to be decreased in patients with different types of calcium mutations causing hypoPP and proposed to be a marker for carrier status (Links and van der Hoeven, 2000).

Since muscle weakness may be triggered by exercise, it has been proposed to use strong and sustained voluntary contraction as a provocative test for diagnosis (McManis et al., 1986). Surface-recorded muscle responses to supramaximal nerve stimulation are used to monitor muscle membrane activity and are considered to reflect muscle membrane activity. Analysis of the compound motor action potential (CMAP) amplitude before and at various times following long (5min) exercise provides information on changes in the number of active fibers, and on their ability to depolarize and repolarize. A significant decrease in the CMAP amplitude following long-exercise test has been reported in ~70–80% of the patients with periodic paralyses (Kuntzer et al., 2000; McManis et al., 1986).

Patients with known ion channel mutations associated with different forms of periodic paralysis were investigated. Inclusive EMG allowed establishing consistent links between the clinical syndromes and the muscle electrical response to different provocative tests (repeated short exercise, long exercise). In addition, statistical analysis of the results obtained from several patients carrying the same mutation provided evidence for the EMG changes caused by specific ion channel mutations. These observations provide evidence that exercise EMG may guide clinicians toward a specific ion channel gene defect especially if access to genetic screening is limited (Fournier et al., 2004; Michel et al., 2007).

Fournier and collaborators proposed a new classification for EMG patterns in patients with muscle channelopathies (Fournier et al., 2004). Periodic paralysis patients could be divided into two groups defined as patterns IV and V (Fournier et al., 2004). The decline in CMAP response, which occurs 15–20 min after completion of a long exercise, is a common feature to both patterns. This loss of muscle excitability correlates well with muscle weakness experienced by patients after strenuous exercise. An early incremental effect of repeated short exercise or long exercise on CMAPs was specific to patients with hyperPP (pattern IV). Recording of a late CMAP decline after long exercise without preliminary increment (pattern V) is most consistent with mutations in CACNA1S or in SCN4A. A similar pattern has been observed in patients with mutations in KCNJ2 (Bendahhou et al., 2005). This suggests that hypoPP-associated calcium channel mutations and Andersen–Tawil syndrome-associated KCNJ2 mutations may lead to muscle membrane hypoexcitability through a common mechanism.

The development of modern imagery has been applied to periodic paralysis. Sodium accumulation in muscles was indeed recently demonstrated using an MRI (Weber et al., 2006). This observation confirms what was inferred was measurements of ion balance during attacks of PP.

VI. TREATMENT AND CARE OF hypoPP Patients

Patients learn to decrease the number of attacks by having a balanced diet: to avoid fasting in hyperPP and carbohydrates in hypoPP. Mild and regular exercise is also beneficial, and continuing to mildly exercise may help abort impending attacks. The use of potassium chloride salts may also be helpful to prevent or abort attacks in hypoPP. Based on the simple reasoning that diuretics inducing hypokalemia may be useful in hyperkalemic periodic paralysis, acetazolamide, an inhibitor of carbonic anhydrase, was tested with success by McArdle (Buruma and Schipperheyn, 1979). A patient suspected to have the hyperkalemic form of periodic paralysis but eventually diagnosed as having hypoPP unexpectedly reported a remarkable beneficial effect of acetazolamide on attack frequency (Resnick *et al.*, 1968). The mechanism of action of acetazolamide is probably more complex that previously thought and still largely unknown. It may be effective in hyperPP through the acidification of the intracellular compartment, thereby slowing all ion channel kinetics in a nonspecific manner. In hypoPP, its action could be mediated by an effect on potassium channels implicated in regulating muscle excitability (Tricarico *et al.*, 2000). In addition, acetazolamide also improves muscle force when permanent weakness is present. The efficacy of carbonic anhydrase inhibitors such as acetazolamide or dichlorphenamide on the frequency of attacks has been demonstrated not only in small series of patients, but also in a multicenter, double-blind, randomized, placebo-controlled cross-over trial (Tawil *et al.*, 2000). An international multicenter trial coordinated by R. C. Griggs (University of Rochester, Rochester, NY) is underway to test the efficacy of carbonic anhydrase inhibitors, not only on the frequency of attacks but also on permanent muscle weakness. It may help to assess whether carbonic anhydrase inhibitors are effective in preventing the occurrence of myopathy.

Some patients with hypoPP worsen with the use of acetazolamide (Torres *et al.*, 1981). It is now known that most of these patients display a sodium channel mutation (Sternberg *et al.*, 2001). However, on an individual basis, it is impossible to predict the response of a patient to acetazolamide, even when knowing the causative mutation (Venance *et al.*, 2004). Although a majority of patients with sodium channel mutations do not respond or are aggravated by acetazolamide, some patients still respond positively. The drug therefore merits a trial under strict medical supervision. Mechanisms that could explain such a differential therapeutic response in patients bearing the same mutations are not known.

There are some anecdotal reports of the association of PP and malignant hyperthermia. The question of the association of PP and malignant hyperthermia arose from the difficulty of interpreting unambiguously *in vitro* contracture tests in muscle disorders caused by ion channel mutations (Lehmann-Horn and Iaizzo, 1990). We now know that malignant hyperthermia and PP are different diseases caused, at least for hypoPP, by distinct mutations in the same gene, the voltage-gated calcium channel *CACNA1S* (Hogan, 1998; Monnier *et al.*, 1997). Some precautions should be taken during anesthetic procedures in hypoPP patients: intravenous glucose should be avoided, and the temperature of fluids and their ionic composition should be carefully monitored. Spinal anesthesia in hypoPP is safe although a known cause of hypokalemia (Hecht *et al.*, 1997). In hyperPP, depolarization agents should be used with caution since they may aggravate myotonia which when it affects jaw muscles or the diaphragm may interfere with ventilation of the anesthetized patients. The balance of ions should be monitored carefully and warm fluids preferred, to avoid paralysis at wakeup.

VII. CONCLUSION

In the past 15years, the combination of advances in genetics, molecular, and clinical physiology has greatly improved our understanding of PP. It now gives a rationale to treatment. The complex links between an ion channel mutation and a phenotype, so complex still remains to be unraveled. Ion channel mutations not only are indeed responsible for a functional defect of the muscle membrane, but also cause muscle degeneration. PP may therefore be good models to decipher the pathway connecting an ion channel mutation to cell toxicity, such as they have been good models to establish the concept of channelopathy. The collaboration of referral centers on an international basis will allow the gathering of data from large numbers of patients which are necessary to accurately describe the natural history of these disorders, a necessary step to conduct therapeutic trials in PP.

Acknowledgments

I thank Emmanuel Fournier, Damien Sternberg, Nacira Tabti, Lucette Lacomblez, Savine Vicart, and Marianne Arzel-Hézode for their active contribution to the work of our group. I thank the members of the French clinical and research network Résocanaux for their active collaborations and discussions (Drs. and Profs. Bassez, Bourdain, Calvas, Chapon, Desnuelle, Eymard, Ferrer, Furby, Gouider, Kuntzer, Labarre-Vila, Lacour, Laforet, Lagueny, Magy, Nicole, Penisson-Besnier, Pereon, Pouget, Sacconi, Tranchant, Vallat, Vermersch, Verschueren, and Vial). The work of the author is supported by Agence Nationale de la Recherche (ANR-Programme Maladies Rares), Association Française contre les Myopathies (AFM), and Institut National de la Recherche Médicale (INSERM).

References

Abbott, G. W., Butler, M. H., Bendahhou, S., Dalakas, M. C., Ptacek, L. J., and Goldstein, S. A. (2001). MiRP2 forms potassium channels in skeletal muscle with Kv3.4 and is associated with periodic paralysis. *Cell* **104**(2), 217–231.

Bendahhou, S., Fournier, E., Sternberg, D., Bassez, G., Furby, A., Sereni, C., Donaldson, M. R., Larroque, M. M., Fontaine, B., and Barhanin, J. (2005). In vivo and in vitro functional characterization of Andersen's syndrome mutations. J. Physiol. 565(Pt. 3), 731–741.

Bradley, W. G., Taylor, R., Rice, D. R., Hausmanowa-Petruzewicz, I., Adelman, L. S., Jenkison, M., Jedrzejowska, H., Drac, H., and Pendlebury, W. W. (1990). Progressive myopathy in hyperkalemic periodic paralysis. Arch. Neurol. 47(9), 1013–1017.

Brancati, F., Valente, E. M., Davies, N. P., Sarkozy, A., Sweeney, M. G., LoMonaco, M., Pizzuti, A., Hanna, M. G., and Dallapiccola, B. (2003). Severe infantile hyperkalaemic periodic paralysis and paramyotonia congenita: Broadening the clinical spectrum associated with the T704M mutation in SCN4A. J. Neurol. Neurosurg. Psychiatry 74(9), 1339–1341.

Bulman, D. E., Scoggan, K. A., van Oene, M. D., Nicolle, M. W., Hahn, A. F., Tollar, L. L., and Ebers, G. C. (1999). A novel sodium channel mutation in a family with hypokalemic periodic paralysis. Neurology 53(9), 1932–1936.

Buruma, O. J., and Bots, G. T. (1978). Myopathy in familial hypokalaemic periodic paralysis independent of paralytic attacks. Acta Neurol. Scand. 57(2), 171–179.

Buruma, O. J. S., and Schipperheyn, J. J. (1979). Periodic paralysis. In "Handbook of Clinical Neurology" (P. J. Vinken and G. W. Bruyn, eds.), pp. 147–174. North-Holland Publishing Company, Amsterdam.

Caciotti, A., Morrone, A., Domenici, R., Donati, M. A., and Zammarchi, E. (2003). Severe prognosis in a large family with hypokalemic periodic paralysis. Muscle Nerve 27(2), 165–169.

Cannon, S. C. (2006). Pathomechanisms in channelopathies of skeletal muscle and brain. Annu. Rev. Neurosci. 29, 387–415.

Cannon, S. C., and Strittmatter, S. M. (1993). Functional expression of sodium channel mutations identified in families with periodic paralysis. Neuron 10(2), 317–326.

Cannon, S. C., Brown, R. H., Jr, and Corey, D. P. (1993). Theoretical reconstruction of myotonia and paralysis caused by incomplete inactivation of sodium channels. Biophys. J. 65(1), 270–288.

Carle, T., Lhuillier, L., Luce, S., Sternberg, D., Devuyst, O., Fontaine, B., and Tabti, N. (2006). Gating defects of a novel Na(+) channel mutant causing hypokalemic periodic paralysis. Biochem. Biophys. Res. Commun. 348(2), 653–661.

Chinnery, P. F., Walls, T. J., Hanna, M. G., Bates, D., and Fawcett, P. R. (2002). Normokalemic periodic paralysis revisited: Does it exist? Ann. Neurol. 52(2), 251–252.

Cummins, T. R., and Sigworth, F. J. (1996). Impaired slow inactivation in mutant sodium channels. Biophys. J. 71(1), 227–236.

Dalakas, M. C., and Engel, W. K. (1983). Treatment of "permanent" muscle weakness in familial Hypokalemic Periodic Paralysis. Muscle Nerve 6(3), 182–186.

Davies, N. P., Sutton, I., Winer, J. B., Moorcroft, P., Pall, H. S., Cole, T., Davies, M. B., Valente, E. M., Brancati, F., Hammans, S. R., and Hanna, M. G. (2002). The sodium channel syndromes: Expanding the phenotype associated with SCN4A mutations. J. Neurol. Neurosurg. Psychiatry 73, 229.

Dias Da Silva, M. R., Cerutti, J. M., Arnaldi, L. A., and Maciel, R. M. (2002). A mutation in the KCNE3 potassium channel gene is associated with susceptibility to thyrotoxic hypokalemic periodic paralysis. J. Clin. Endocrinol. Metab. 87(11), 4881–4884.

Elbaz, A., Vale-Santos, J., Jurkat-Rott, K., Lapie, P., Ophoff, R. A., Bady, B., Links, T. P., Piussan, C., Vila, A., Monnier, N., et al. (1995). Hypokalemic periodic paralysis and the dihydropyridine receptor (CACNL1A3): Genotype/phenotype correlations for two predominant mutations and evidence for the absence of a founder effect in 16 Caucasian families. Am. J. Hum. Genet. 56(2), 374–380.

Engel, A. G. (1966). Electron microscopic observations in primary hypokalemic and thyrotoxic periodic paralyses. Mayo Clin. Proc. 41(11), 797–808.

Engel, A. G. (1970). Evolution and content of vacuoles in primary hypokalemic periodic paralysis. Mayo Clin. Proc. 45(11), 774–814.

Engel, A. G., and Lambert, E. H. (1969). Calcium activation of electrically inexcitable muscle fibers in primary hypokalemic periodic paralysis. *Neurology* **19**(9), 851–858.

Feero, W. G., Wang, J., Barany, F., Zhou, J., Todorovic, S. M., Conwit, R., Galloway, G., Hausmanowa-Petrusewicz, I., Fidzianska, A., Arahata, K., *et al.* (1993). Hyperkalemic periodic paralysis: Rapid molecular diagnosis and relationship of genotype to phenotype in 12 families. *Neurology* **43**(4), 668–673.

Fontaine, B., Khurana, T. S., Hoffman, E. P., Bruns, G. A., Haines, J. L., Trofatter, J. A., Hanson, M. P., Rich, J., McFarlane, H., Yasek, D. M., *et al.* (1990). Hyperkalemic periodic paralysis and the adult muscle sodium channel alpha-subunit gene. *Science* **250**(4983), 1000–1002.

Fontaine, B., Vale-Santos, J., Jurkat-Rott, K., Reboul, J., Plassart, E., Rime, C., S., Elbaz, A., Heine, R., Guimaraes, J., Weissenbach, J., *et al.* (1994). Mapping of the hypokalaemic periodic paralysis (HypoPP) locus to chromosome 1q31–32 in three European families. *Nat. Genet.* **6**(3), 267–272.

Fontaine, B., Fournier, E., Sternberg, D., Vicart, S., and Tabti, N. (2007). Hypokalemic periodic paralysis: A model for a clinical and research approach to a rare disorder. *Neurotherapeutics* **4**(2), 225–232.

Fouad, G., Dalakas, M., Servidei, S., Mendell, J. R., Van den Bergh, P., Angelini, C., Alderson, K., Griggs, R. C., Tawil, R., Gregg, R., Hogan, K., and Powers, P. A. *et al.* (1997). Genotype–phenotype correlations of DHP receptor alpha 1-subunit gene mutations causing hypokalemic periodic paralysis. *Neuromuscul. Disord.* **7**(1), 33–38.

Fournier, E., Arzel, M., Sternberg, D., Vicart, S., Laforet, P., Eymard, B., Willer, J. C., Tabti, N., and Fontaine, B. (2004). Electromyography guides toward subgroups of mutations in muscle channelopathies. *Ann. Neurol.* **56**(5), 650–661.

Griggs, R. C., Engel, W. K., and Resnick, J. S. (1970). Acetazolamide treatment of hypokalemic periodic paralysis. Prevention of attacks and improvement of persistent weakness. *Ann. Intern. Med.* **73**(1), 39–48.

Hayward, L. J., Sandoval, G. M., and Cannon, S. C. (1999). Defective slow inactivation of sodium channels contributes to familial periodic paralysis. *Neurology* **52**(7), 1447–1453.

Hecht, M. L., Valtysson, B., and Hogan, K. (1997). Spinal anesthesia for a patient with a calcium channel mutation causing hypokalemic periodic paralysis. *Anesth. Analg.* **84**(2), 461–464.

Hogan, K. (1998). The anesthetic myopathies and malignant hyperthermias. *Curr. Opin. Neurol.* **11**(5), 469–476.

Ikeda, Y., Abe, B., Watanabe, M., Shoji, M., Fontaine, B., Itoyama, Y., and Hirai, S. (1996). A Japanese family of autosomal dominant hypokalemic periodic paralysis with a CACNL1A3 gene mutation. *Eur. J. Neurol.* **3**, 441–445.

Jurkat-Rott, K., and Lehmann-Horn, F. (2004). Periodic paralysis mutation MiRP2-R83H in controls: Interpretations and general recommendation. *Neurology* **62**(6), 1012–1015.

Jurkat-Rott, K., and Lehmann-Horn, F. (2007). Genotype–phenotype correlation and therapeutic rationale in hyperkalemic periodic paralysis. *Neurotherapeutics* **4**(2), 216–224.

Jurkat-Rott, K., Lehmann-Horn, F., Elbaz, A., Heine, R., Gregg, R. G., Hogan, K., Powers, P. A., Lapie, P., Vale-Santos, J. E., Weissenbach, J., *et al.* (1994). A calcium channel mutation causing hypokalemic periodic paralysis. *Hum. Mol. Genet.* **3**(8), 1415–1419.

Jurkat-Rott, K., Mitrovic, N., Hang, C., Kouzmekine, A., Iaizzo, P., Herzog, J., Lerche, H., Nicole, S., Vale-Santos, J., Chauveau, D., Fontaine, B., and Lehmann-Horn, F. (2000). Voltage-sensor sodium channel mutations cause hypokalemic periodic paralysis type 2 by enhanced inactivation and reduced current. *Proc. Natl Acad. Sci. USA* **97**(17), 9549–9554.

Kawamura, S., Ikeda, Y., Tomita, K., Watanabe, N., and Seki, K. (2004). A family of hypokalemic periodic paralysis with CACNA1S gene mutation showing incomplete penetrance in women. *Intern. Med.* **43**(3), 218–222.

Kelly, P., Yang, W. S., Costigan, D., Farrell, M. A., Murphy, S., and Hardiman, O. (1997). Paramyotonia congenita and hyperkalemic periodic paralysis associated with a Met 1592 Val substitution in the skeletal muscle sodium channel alpha subunit—a large kindred with a novel phenotype. *Neuromuscul. Disord.* **7**(2), 105–111.

Kim, J., Hahn, Y., Sohn, E. H., Lee, Y. J., Yun, J. H., Kim, J. M., and Chung, J. H. (2001). Phenotypic variation of a Thr704Met mutation in skeletal sodium channel gene in a family with paralysis periodica paramyotonica. *J. Neurol. Neurosurg. Psychiatry* **70**(5), 618–623.

Kung, A. W. (2006). Clinical review: Thyrotoxic periodic paralysis: A diagnostic challenge. *J. Clin. Endocrinol. Metab.* **91**(7), 2490–2495.

Kuntzer, T., Flocard, F., Vial, C., Kohler, A., Magistris, M., Labarre-Vila, A., Gonnaud, P. M., Ochsner, F., Soichot, P., Chan, V., and Monnier, G. (2000). Exercise test in muscle channelopathies and other muscle disorders. *Muscle Nerve* **23**(7), 1089–1094.

Lapie, P., Goudet, C., Nargeot, J., Fontaine, B., and Lory, P. (1996). Electrophysiological properties of the hypokalaemic periodic paralysis mutation (R528H) of the skeletal muscle alpha 1s subunit as expressed in mouse L cells. *FEBS Lett.* **382**(3), 244–248.

Lehmann-Horn, F., and Iaizzo, P. A. (1990). Are myotonias and periodic paralyses associated with susceptibility to malignant hyperthermia? *Br. J. Anaesth.* **65**(5), 692–697.

Lehmann-Horn, F., Rudel, R., and Ricker, K. (1993). Non-dystrophic myotonias and periodic paralyses. A European Neuromuscular Center Workshop held 4–6 October 1992, Ulm, Germany. *Neuromuscul. Disord.* **3**(2), 161–168.

Links, T. P. (1992). *In* "Familial Hypokalemic Periodic Paralysis." p. 175. Rijksuniversiteit Groningen, Groningen.

Links, T. P., and van der Hoeven, J. H. (2000). Muscle fiber conduction velocity in arg1239his mutation in hypokalemic periodic paralysis. *Muscle Nerve* **23**(2), 296.

Links, T. P., Zwarts, M. J., and Oosterhuis, H. J. (1988). Improvement of muscle strength in familial hypokalaemic periodic paralysis with acetazolamide. *J. Neurol. Neurosurg. Psychiatry* **51**(9), 1142–1145.

Links, T. P., Zwarts, M. J., Wilmink, J. T., Molenaar, W. M., and Oosterhuis, H. J. (1990). Permanent muscle weakness in familial hypokalaemic periodic paralysis. Clinical, radiological and pathological aspects. *Brain* **113**(Pt. 6), 1873–1889.

Links, T. P., Smit, A. J., Molenaar, W. M., Zwarts, M. J., and Oosterhuis, H. J. (1994a). Familial hypokalemic periodic paralysis. Clinical, diagnostic and therapeutic aspects. *J. Neurol. Sci.* **122**(1), 33–43.

Links, T. P., van der Hoeven, J. H., and Zwarts, M. J. (1994b). Surface EMG and muscle fibre conduction during attacks of hypokalaemic periodic paralysis. *J. Neurol. Neurosurg. Psychiatry* **57**(5), 632–634.

McManis, P. G., Lambert, E. H., and Daube, J. R. (1986). The exercise test in periodic paralysis. *Muscle Nerve* **9**(8), 704–710.

Michel, P., Sternberg, D., Jeannet, P. Y., Dunand, M., Thonney, F., Kress, W., Fontaine, B., Fournier, E., and Kuntzer, T. (2007). Comparative efficacy of repetitive nerve stimulation, exercise, and cold in differentiating myotonic disorders. *Muscle Nerve* **36**(5), 643–650.

Miller, T. M., Dias Da Silva, M. R., Miller, H. A., Kwiecinski, H., Mendell, J. R., Tawil, R., McManis, P., Griggs, R. C., Angelini, C., Servidei, S., Petajan, J., and Dalakas, M. C. *et al.* (2004). Correlating phenotype and genotype in the periodic paralyses. *Neurology* **63**(9), 1647–1655.

Monnier, N., Procaccio, V., Stieglitz, P., and Lunardi, J. (1997). Malignant-hyperthermia susceptibility is associated with a mutation of the alpha 1-subunit of the human dihydropyridine-sensitive L-type voltage-dependent calcium-channel receptor in skeletal muscle. *Am. J. Hum. Genet.* **60**(6), 1316–1325.

Morrill, J. A., and Cannon, S. C. (1999). Effects of mutations causing hypokalaemic periodic paralysis on the skeletal muscle L-type Ca2+ channel expressed in *Xenopus laevis* oocytes. *J. Physiol.* **520**(Pt. 2), 321–336.

Okuda, S., Kanda, F., Nishimoto, K., Sasaki, R., and Chihara, K. (2001). Hyperkalemic periodic paralysis and paramyotonia congenita—a novel sodium channel mutation. *J. Neurol.* **248**(11), 1003–1004.

Plassart, E., Reboul, J., Rime, C. S., Recan, D., Millasseau, P., Eymard, B., Pelletier, J., Thomas, C., Chapon, F., Desnuelle, C., *et al.* (1994). Mutations in the muscle sodium channel gene (SCN4A) in 13 French families with hyperkalemic periodic paralysis and paramyotonia congenita: Phenotype to genotype correlations and demonstration of the predominance of two mutations. *Eur. J. Hum. Genet.* **2**(2), 110–124.

Plassart, E., Eymard, B., Maurs, L., Hauw, J. J., Lyon-Caen, O., Fardeau, M., and Fontaine, B. (1996). Paramyotonia congenita: Genotype to phenotype correlations in two families and report of a new mutation in the sodium channel gene. *J. Neurol. Sci.* **142**(1–2), 126–133.

Plaster, N. M., Tawil, R., Tristani-Firouzi, M., Canun, S., Bendahhou, S., Tsunoda, A., Donaldson, M. R., Iannaccone, S. T., Brunt, E., Barohn, R., Clark, J., and Deymeer, F. *et al.* (2001). Mutations in Kir2.1 cause the developmental and episodic electrical phenotypes of Andersen's syndrome. *Cell* **105**(4), 511–519.

Ptacek, L. J., George, A. L., Jr, Griggs, R. C., Tawil, R., Kallen, R. G., Barchi, R. L., Robertson, M., and Leppert, M. F. (1991). Identification of a mutation in the gene causing hyperkalemic periodic paralysis. *Cell* **67**(5), 1021–1027.

Ptacek, L. J., Gouw, L., Kwiecinski, H., McManis, P., Mendell, J. R., Barohn, R. J., George, A. L., Jr, Barchi, R. L., Robertson, M., and Leppert, M. F. (1993). Sodium channel mutations in paramyotonia congenita and hyperkalemic periodic paralysis. *Ann. Neurol.* **33**(3), 300–307.

Ptacek, L. J., Tawil, R., Griggs, R. C., Engel, A. G., Layzer, R. B., Kwiecinski, H., McManis, P. G., Santiago, L., Moore, M., Fouad, G., *et al.* (1994). Dihydropyridine receptor mutations cause hypokalemic periodic paralysis. *Cell* **77**(6), 863–868.

Resnick, J. S., Engel, W. K., Griggs, R. C., and Stam, A. C. (1968). Acetazolamide prophylaxis in hypokalemic periodic paralysis. *N. Engl. J. Med.* **278**(11), 582–586.

Ricker, K., Rohkamm, R., and Bohlen, R. (1986). Adynamia episodica and paralysis periodica paramyotonica. *Neurology* **36**(5), 682–686.

Rojas, C. V., Wang, J. Z., Schwartz, L. S., Hoffman, E. P., Powell, B. R., and Brown, R. H., Jr (1991). A Met-to-Val mutation in the skeletal muscle Na+ channel alpha-subunit in hyperkalaemic periodic paralysis. *Nature* **354**(6352), 387–389.

Rudel, R., Lehmann-Horn, F., Ricker, K., and Kuther, G. (1984). Hypokalemic periodic paralysis: *In vitro* investigation of muscle fiber membrane parameters. *Muscle Nerve* **7**(2), 110–120.

Ruff, R. L. (1991). Calcium–tension relationships of muscle fibers from patients with periodic paralysis. *Muscle Nerve* **14**(9), 838–844.

Ruff, R. L. (1999). Insulin acts in hypokalemic periodic paralysis by reducing inward rectifier K+ current. *Neurology* **53**(7), 1556–1563.

Ruff, R. L. (2008). Slow inactivation: Slow but not dull. *Neurology* **70**(10), 746–747.

Sillen, A., Sorensen, T., Kantola, I., Friis, M. L., Gustavson, K. H., and Wadelius, C. (1997). Identification of mutations in the CACNL1A3 gene in 13 families of Scandinavian origin having hypokalemic periodic paralysis and evidence of a founder effect in Danish families. *Am. J. Med. Genet.* **69**(1), 102–106.

Sternberg, D., Maisonobe, T., Jurkat-Rott, K., Nicole, S., Launay, E., Chauveau, D., Tabti, N., Lehmann-Horn, F., Hainque, B., and Fontaine, B. (2001). Hypokalaemic periodic paralysis type 2 caused by mutations at codon 672 in the muscle sodium channel gene SCN4A. *Brain* **124**(Pt. 6), 1091–1099.

Sternberg, D., Tabti, N., Fournier, E., Hainque, B., and Fontaine, B. (2003). Lack of association of the potassium channel-associated peptide MiRP2-R83H variant with periodic paralysis. *Neurology* **61**(6), 857–859.

Sugiura, Y., Aoki, T., Sugiyama, Y., Hida, C., Ogata, M., and Yamamoto, T. (2000). Temperature-sensitive sodium channelopathy with heat-induced myotonia and cold-induced paralysis. *Neurology* **54**(11), 2179–2181.

Tang, N. L., Chow, C. C., Ko, G. T., Tai, M. H., Kwok, R., Yao, X. Q., and Cockram, C. S. (2004). No mutation in the KCNE3 potassium channel gene in Chinese thyrotoxic hypokalaemic periodic paralysis patients. *Clin. Endocrinol. (Oxf.)* **61**(1), 109–112.

Tawil, R., Ptacek, L. J., Pavlakis, S. G., DeVivo, D. C., Penn, A. S., Ozdemir, C., and Griggs, R. C. (1994). Andersen's syndrome: Potassium-sensitive periodic paralysis, ventricular ectopy, and dysmorphic features. *Ann. Neurol.* **35**(3), 326–330.

Tawil, R., McDermott, M. P., Brown, R., Jr, Shapiro, B. C., Ptacek, L. J., McManis, P. G., Dalakas, M. C., Spector, S. A., Mendell, J. R., Hahn, A. F., and Griggs, R. C. (2000). Randomized trials of dichlorphenamide in the periodic paralyses. Working Group on Periodic Paralysis. *Ann. Neurol.* **47**(1), 46–53.

Torres, C. F., Griggs, R. C., Moxley, R. T., and Bender, A. N. (1981). Hypokalemic periodic paralysis exacerbated by acetazolamide. *Neurology* **31**(11), 1423–1428.

Tricarico, D., Servidei, S., Tonali, P., Jurkat-Rott, K., and Camerino, D. C. (1999). Impairment of skeletal muscle adenosine triphosphate-sensitive K+ channels in patients with hypokalemic periodic paralysis. *J. Clin. Invest.* **103**(5), 675–682.

Tricarico, D., Barbieri, M., and Camerino, D. C. (2000). Acetazolamide opens the muscular KCa2+ channel: A novel mechanism of action that may explain the therapeutic effect of the drug in hypokalemic periodic paralysis. *Ann. Neurol.* **48**(3), 304–312.

Venance, S. L., Jurkat-Rott, K., Lehmann-Horn, F., and Tawil, R. (2004). SCN4A-associated hypokalemic periodic paralysis merits a trial of acetazolamide. *Neurology* **63**(10), 1977.

Venance, S. L., Cannon, S. C., Fialho, D., Fontaine, B., Hanna, M. G., Ptacek, L. J., Tristani-Firouzi, M., Tawil, R., and Griggs, R. C. (2006). The primary periodic paralyses: Diagnosis, pathogenesis and treatment. *Brain* **129**(Pt. 1), 8–17.

Vicart, S., Sternberg, D., Fournier, E., Ochsner, F., Laforet, P., Kuntzer, T., Eymard, B., Hainque, B., and Fontaine, B. (2004). New mutations of SCN4A cause a potassium-sensitive normokalemic periodic paralysis. *Neurology* **63**(11), 2120–2127.

Wagner, S., Lerche, H., Mitrovic, N., Heine, R., George, A. L., and Lehmann-Horn, F. (1997). A novel sodium channel mutation causing a hyperkalemic paralytic and paramyotonic syndrome with variable clinical expressivity. *Neurology* **49**(4), 1018–1025.

Weber, M. A., Nielles-Vallespin, S., Essig, M., Jurkat-Rott, K., Kauczor, H. U., and Lehmann-Horn, F. (2006). Muscle Na+ channelopathies: MRI detects intracellular 23Na accumulation during episodic weakness. *Neurology* **67**(7), 1151–1158.

Zwarts, M. J., van Weerden, T. W., Links, T. P., Haenen, H. T., and Oosterhuis, H. J. (1988). The muscle fiber conduction velocity and power spectra in familial hypokalemic periodic paralysis. *Muscle Nerve* **11**(2), 166–173.

2

Myotonia Congenita

Christoph Lossin* and Alfred L. George, Jr.[†]

*Department of Neurology, UC Davis School of Medicine, Sacramento, California 95817
[†]Departments of Medicine and Pharmacology, Vanderbilt University School of Medicine, Nashville, Tennessee 37232

I. Introduction
II. Clinical Aspects of Myotonia Congenita
    A. Autosomal-dominant myotonia congenita (Thomsen Disease)
    B. Recessive generalized myotonia (Becker myotonia)
    C. Diagnosis of myotonia congenita
III. Molecular Genetics of Myotonia Congenita
    A. Muscle chloride channel ClC-1
    B. Spectrum of CLCN1 mutations in myotonia congenita
IV. Physiological Basis of Myotonia Congenita
    A. Role of the sarcolemmal chloride conductance
    B. Functional consequences of CLCN1 mutations
    C. Unanswered questions
V. Myotonia Congenita in Animal Models
VI. Treatment of Myotonia Congenita
    A. Pharmacological therapies
    B. Other therapeutic considerations
    C. Gene therapy for myotonia congenita
VII. Concluding Remarks
    Acknowledgments
    References

Advances in Genetics, Vol. 63
Copyright 2008, Elsevier Inc. All rights reserved.

0065-2660/08 $35.00
DOI: 10.1016/S0065-2660(08)01002-X

ABSTRACT

Myotonia is a symptom of many different acquired and genetic muscular conditions that impair the relaxation phase of muscular contraction. Myotonia congenita is a specific inherited disorder of muscle membrane hyperexcitability caused by reduced sarcolemmal chloride conductance due to mutations in *CLCN1*, the gene coding for the main skeletal muscle chloride channel ClC-1. The disorder may be transmitted as either an autosomal-dominant or recessive trait with close to 130 currently known mutations. Although this is a rare disorder, elucidation of the pathophysiology underlying myotonia congenita established the importance of sarcolemmal chloride conductance in the control of muscle excitability and demonstrated the first example of human disease associated with the ClC family of chloride transporting proteins. © 2008, Elsevier Inc.

I. INTRODUCTION

This chapter reviews the clinical aspects, genetics, molecular pathophysiology, and treatment of myotonia congenita. In this disorder, impaired functioning of the skeletal muscle chloride channel (ClC-1) leads to an increase in sarcolemmal excitability that clinically presents as delayed muscular relaxation (myotonia). The culprit for this disorder, deficient muscle chloride conductance, emerged from seminal studies by Bryant, Lipicky, and colleagues (Bryant, 1962; Bryant and Morales-Aguilera, 1971; Lipicky and Bryant, 1966; Morgan *et al.*, 1975). In the last 30 years, with the extraordinary accomplishments and discoveries of modern molecular biology, these pioneering findings were substantiated and we now have direct insight into the pathophysiology of this condition. Today, many of the ion channel molecules that participate in generating action potentials have been defined at the primary nucleotide sequence level, and this work has paved the way toward the genetic basis of many excitability disorders highlighted in other chapters. In parallel, advances in cellular electrophysiology coupled with the use of recombinant ion channels have enabled us to expand our knowledge of the molecular pathophysiology of such disorders. Myotonia congenita, although a rare disorder, provided important clues as to the importance of sarcolemmal chloride conductance in controlling muscle membrane excitability.

II. CLINICAL ASPECTS OF MYOTONIA CONGENITA

Myotonia (*myo-* from Greek: muscle; and *tonus* from Latin: tension) describes an abnormal delay in muscle relaxation following voluntary forceful contraction (Gutmann and Phillips, 1991; Streib, 1987). Affected individuals describe muscular stiffness upon initiating movement. The stiffness remits with several repetitions of the same movement, giving rise to the so-called *warm-up* phenomenon. In some cases, the myotonia does not begin until one or two sets of the same movement are completed (*delayed myotonia*). If symptoms worsen with repeated exercise, the term *paradoxical myotonia* is more appropriate, but this tends to occur in the setting of mutant muscle sodium channels rather than in myotonia congenita. Myotonia may occur with a variety of other disorders including hyperkalemic periodic paralysis, paramyotonia congenita, potassium-aggravated myotonia, and myotonic muscular dystrophy. Exposure to certain drugs and chemical toxins can also induce myotonia. The clinical presentation of these conditions overlaps in some aspects, but a thorough review of symptoms, clinical history, and inheritance pattern usually permits a clear differential diagnosis (Table 2.1).

Myotonia congenita may be inherited as either an autosomal-dominant (Thomsen Disease, OMIM 160800) or recessive trait (recessive generalized myotonia or Becker myotonia, OMIM 255700). The clinical features of both conditions are very similar but can be distinguished by severity and inheritance pattern. Both disorders are caused by mutations in the gene (*CLCN1*) encoding the skeletal muscle chloride channel ClC-1.

A. Autosomal-dominant myotonia congenita (Thomsen Disease)

The first thorough description of myotonia congenita dates back to 1876, when Julius Thomsen, a physician from the Schleswig region of North Germany, published a manuscript on a muscle ailment afflicting various members of his family including himself (Thomsen, 1876). The term *Thomsen Disease* was later coined (Westphal, 1883).

Table 2.1. Clinical Presentations of Myotonia Congenita

	Recessive generalized myotonia	Thomsen Disease
Inheritance	Recessive	Dominant
Onset	4–6 yrs or younger	Birth
Muscles involved	Gradual onset but generalizing to involve all muscle	Individual muscle groups
Muscle strength	Normal after remission from transient weakness	Normal to supernormal
Severity	Moderate to severe	Mild to moderate
Muscular hypertrophy	Mild to obvious, however atrophy in forearms	None to mild
Gender ratio	$\female \approx \male$	$\female \approx \male$[a]

[a]Some data in the literature suggest a slight predominance of male over female myotonia. Traditional culture differences and gender roles may have biased the acquisition of these data.

Dominantly inherited myotonia congenita is usually evident during early infancy (Becker, 1977). The first symptom may be delayed relaxation of the eyelids after forceful closure following sneezing or during crying (von Graefe's sign or lid lag; Wakeman et al., 2008). The infant may also present with unusually defined muscles in the extremities, but overt muscular hypertrophy in infancy is rare. Frequently, Thomsen Disease is not recognized until late childhood, when the child appears clumsy and has difficulties with movements after rest. The muscles are stiff, especially in the legs, upon rising from a seated position, and hesitancy during stair climbing may be evident.

Thomsen Disease can vary in severity from mild to moderate with severe cases being rare. Life expectancy is normal and affected individuals can lead a satisfying and successful life. Genetic selection against Thomsen Disease is nonexistent (Becker, 1977). There is no progression of the symptoms and most affected subjects learn to live with the condition. Although, in severe cases of Thomsen Disease and recessive generalized myotonia (RGM, see below), there can be elevated risk of trauma as the ability for self-righting following loss of balance is compromised (Freeble and Rini, 1949). There are no central nervous system manifestations. However, a psychiatric burden can be evident, and especially in young males where muscular hypertrophy is obvious, mockery and misunderstanding by peers on account of poor physical performance and "clumsiness" can be problematic (Becker, 1977). Invariably, individuals feel uncomfortable about revealing their condition and commonly develop highly effective measures to mask the malady. In some cases, the desire to evade the condition can become strong enough for the affected individual to consider empirical self-medication with drugs obtained from nonmedical suppliers that may pose other health risks.

Thomsen Disease is dominantly inherited, but pronounced intrafamilial heterogeneity is not uncommon (Chang et al., 2007). Age of onset, affected muscle groups, and severity can differ significantly between siblings or children and parents (reviewed in Colding-Jorgensen, 2005). On occasion, the mutation carriers' physical performance is only marginally affected; some may in fact be unaware of their condition or be clinically asymptomatic. Electromyographic evaluation, however, may demonstrate evidence of abnormal discharges (*latent myotonia*).

B. Recessive generalized myotonia (Becker myotonia)

RGM is similar to Thomsen Disease except for recessive inheritance and a generally more severe phenotype. The eponym for this disorder is in honor of the German Professor Peter Emil Becker who devoted many years of his life personally examining patients with congenital myotonia, recognizing that there was a subset that was genetically distinct from Thomsen Disease (Becker, 1966).

Clinically, RGM presents in early childhood or infancy with generalized myotonia accompanied by moderate to pronounced muscular hypertrophy that presumably develops because of chronically increased muscle activity. Compared with Thomsen Disease, individuals afflicted with RGM have more severe myotonia and may experience transient periods (<1 min) of weakness upon initiating movement, especially after prolonged rest (Rüdel et al., 1988). Transient weakness manifests suddenly with the perception of a "wave" of weakness or muscle failure affecting muscles involved in a task and possibly results in the dropping of an object or impaired postural control. Severe myotonia may compromise postural control to the degree that righting movements are prematurely "frozen" in mid-action leading to uncontrolled falling. Although myotonia is normally not associated with pain, subjects with RGM may infrequently develop painful muscle cramps or myalgia (Becker, 1977), particularly during rest following vigorous exercise. Despite generalized muscular hypertrophy, some individuals with RGM may have atrophy of the forearms (Becker, 1977). Electromyographic analyses on the upper extremities should therefore be avoided. Reduced upper limb reflexes (Becker, 1977; Fialho et al., 2007) and limited wrist dorsiflexion (Becker, 1977) have also been observed.

In general, women appear to be affected less often than men (Becker, 1966), but bias in data ascertainment may skew the gender ratio. Temporary fluctuations in myotonia severity may occur in response to dietary insufficiencies, sleep deprivation, prolonged physical activity, and emotional stress. Anecdotally, pregnancy and hypothyroidism may alleviate or worsen symptoms in some affected individuals. Similarly, latent myotonia may become clinically apparent during menses, which may be the result of sex hormone-mediated functional alterations of CIC-1 (Fialho et al., 2008).

C. Diagnosis of myotonia congenita

Myotonia must be first recognized from the characteristic symptoms, but many physicians are unfamiliar with this clinical phenomenon and may attribute myotonic symptoms to other conditions. A family history of myotonia is helpful for distinguishing Thomsen Disease from RGM in some, but not all, cases owing to variable expressivity. A true autosomal-dominant inheritance pattern excludes a diagnosis of RGM, but absence of myotonia in the parents is less informative because the disorder may arise by de novo mutation.

To test for myotonia at the bedside, a reflex hammer can be used on prominent muscles like the thenar eminence, the side of the tongue, or the thigh (Fig. 2.1), which produces clearly discernible dimpling (percussion myotonia) or fasciculation upon impact that lingers for several seconds. Absence of this characteristic reaction does not rule out myotonia, however, owing to considerable variations in the phenotypic presentation (day to day, between individuals, muscles involved, etc.). Objective evidence for myotonia can be

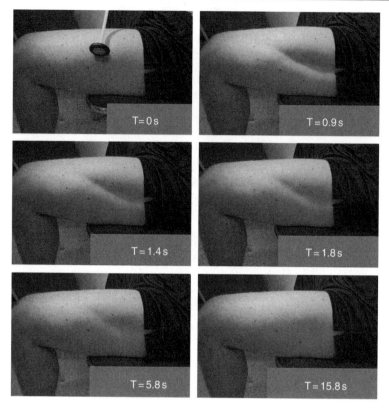

Figure 2.1. Percussion myotonia of the thigh. A reflex hammer was used to tap the distal end of the vastus lateralis and this produced a distinct contraction of the entire muscle that persisted for several seconds. The subject in this case carries a diagnosis of RGM and is heterozygous for a frameshift mutation in *CLCN1* with a presumed undetected second mutation on the opposite allele.

provided by needle electromyography. Myotonia is characterized by increased insertional activity and continued high-frequency discharges that outlast motor unit responses from healthy subjects by several seconds following tapping, needle movement, or voluntary contraction. The audio feedback of myotonia during electromyography is unmistakable, resembling the sound produced by propeller dive bombers during World War II. The duration of repetitive discharges correlates well with delayed muscle relaxation and represents the electrophysiological hallmark of the disorder.

Myotonia congenita may be accompanied by histological changes in muscle. Abnormal variations in myocyte diameter, fiber hypertrophy, absence of type IIb fibers, central location of fiber nuclei, and mitochondrial aggregates have been reported (Engel and Brooke, 1966). None of these changes are specific for myotonia congenita and can also be observed in other inherited nondystrophic myotonias. However, subjects with myotonia may warrant an examination of muscle histology as part of an evaluation for myotonic dystrophy.

The differential diagnosis of myotonia includes other inherited myotonic disorders such as the nondystrophic myotonias and periodic paralyses, myotonic dystrophy, proximal myotonic myopathy (PROMM), drug-induced myotonia and neuromyotonia. In the differential diagnosis of myotonic syndromes, it is imperative to distinguish those associated with dystrophy from the more benign, nondystrophic forms. In most cases, this distinction is easily made on clinical grounds. Myotonic dystrophy, also known as Steinert's disease or myotonia atrophica (OMIM 160900), is a multisystem disorder in which late-onset progressive wasting of bulbar and distal muscles usually predominates over a relatively mild degree of myotonia. Inheritance is dominant and highly penetrant with progressive worsening in successive generations (*genetic anticipation*). Interestingly, the myotonic component stems from aberrant splicing of the muscle chloride channel ClC-1 which is mutated in myotonia congenita (Mankodi *et al.*, 2002).

A small number of drugs and toxins may trigger myotonia by reducing sarcolemmal chloride conductance (Kwiecinski, 1981; Mastaglia, 1982). The best studied agents are aromatic carboxylic acids including 9-anthracene carboxylic acid (9-AC) (Furman and Barchi, 1978), the herbicide 2,4-dichlorophenoxyacetic acid (2,4-D) (Bradberry *et al.*, 2000), and the cholesterol-lowering drug clofibrate (2-(*p*-chlorophenoxy)propionic acid) (Conte-Camerino *et al.*, 1984). These agents produce their clinical effects through distinct mechanisms. Reduction of sarcolemmal chloride conductance by 2,4-D, 9-AC, and several other aromatic carboxylic acid compounds is caused by altered ion selectivity of the muscle chloride channel (Palade and Barchi, 1977). Clofibrate, by contrast, accelerates chloride channel deactivation and shifts the voltage dependence of channel activation to more depolarized potentials (Pusch *et al.*, 2000) similar to many myotonia-producing *CLCN1* mutations (see below). Two drugs used for the treatment of hypercholesterolemia, simvastatin and pravastatin, both HMG-CoA reductase inhibitors, can induce myotonia in rabbits by reducing sarcolemmal chloride conductance (Pierno *et al.*, 1995, 2006; Sonoda *et al.*, 1994).

Certain clinical features may help differentiate Thomsen Disease and RGM from other nondystrophic myotonias. Paradoxical myotonia of the hands, face, and tongue that is aggravated by cooling is highly suggestive of paramyotonia congenita, an inherited disorder caused by mutation of the skeletal muscle sodium channel gene *SCN4A*. Cooling may also worsen symptoms in potassium-aggravated myotonia, whereas Thomsen Disease and RGM are generally not affected by temperature. Age at onset tends to be later in life for myotonic disorders associated with *SCN4A* mutations. Myotonia associated with episodic or persistent weakness is suggestive of other disorders as well (periodic paralysis and potassium-aggravated myotonia). Electromyography is not capable of distinguishing Thomsen Disease and RGM from other myotonic disorders.

An emerging diagnostic tool is molecular genetic testing. Genetic analysis of *CLCN1* (Thomsen Disease and RGM) is now commercially available. With this assay, the detection of a nonsense or frameshift mutation is highly informative, but missense alleles can be problematic unless determined previously to exhibit abnormal function using an *in vitro* assay or there is evidence of cosegregation of the variant with the phenotype in the affected family. The current commercial assay uses DNA sequencing of polymerase chain reaction-amplified exons and flanking intron regions of *CLCN1*, but this method will not detect large segmental gene deletions or duplications. Therefore, a negative screen for exonic mutations will not completely rule out the disorder. Additional screens targeting genes linked to diseases with myotonia as part of the clinical spectrum such as *SCN4A* (paramyotonia congenita, hyperkalemic periodic paralysis, and potassium-aggravated myotonia) as well as *DMPK* and *ZNF9* (myotonic dystrophy, types 1 and 2) may be necessary (Fialho *et al.*, 2007; Trip *et al.*, 2008).

III. MOLECULAR GENETICS OF MYOTONIA CONGENITA

In the 1960s, Bryant and colleagues developed the hypothesis that the pathogenesis of myotonia congenita involved defects of sarcolemmal chloride channel function or regulation (Bryant, 1962; Bryant and Morales-Aguilera, 1971; Lipicky and Bryant, 1966). Approximately 30 years later, the first candidate gene encoding a muscle membrane chloride channel was discovered when Jentsch *et al.* (1990) isolated the complementary DNA sequence of ClC-0, a voltage-sensitive chloride channels in the electric ray, *Torpedo marmorata*. The first mammalian isoform (ClC-1) was identified in rat skeletal muscle (Steinmeyer *et al.*, 1991b).

A. Muscle chloride channel ClC-1

Human ClC-1, a 988 amino acid membrane protein encoded by the 23 exon *CLCN1* gene on chromosome 7q35 (Lorenz *et al.*, 1994), is the principal skeletal muscle chloride channel and was the first appropriate candidate for genetic abnormalities associated with myotonia congenita. Molecular support for this notion came first from the work of Steinmeyer and coworkers in 1991, who found *Clcn1* mutations in homozygous *adr* (_a_rrested _d_evelopment of _r_ighting) mice, an animal model for RGM (Steinmeyer *et al.*, 1991a). One year later, myotonia congenita was genetically mapped close to the *CLCN1* gene in humans

(Abdalla *et al.*, 1992), and soon thereafter the first *CLCN1* mutations for both recessive and autosomal-dominant myotonia congenita were discovered (George *et al.*, 1993; Koch *et al.*, 1992).

The ClC proteins are uniquely constructed with two independently gated pores, analogous to a double-barreled shotgun (Miller and White, 1984). This structural prediction was based originally on detailed analyses of single purified *Torpedo* electroplax Cl$^-$ channels reconstituted in planar lipid bilayers, which demonstrated two equal but independently gated conductance states (Hanke and Miller, 1983; Miller, 1982). Single channel recordings of mammalian ClC-1 and ClC-2 supported this hypothesis (Nobile *et al.*, 2000; Saviane *et al.*, 1999). Additional genetic and biochemical evidence supported a dimeric structure of ClC-1 (Fahlke *et al.*, 1997c; Steinmeyer *et al.*, 1994). In this model, permeation through each subunit is regulated independently by a fast gate, whereas both pores can be simultaneously closed by a common slow gate. Final proof of the double-barreled structure of ClC channels was provided in 2002 by X-ray diffraction experiments on crystallized bacterial ClC proteins that revealed a dimeric architecture with individual pores in each subunit (Dutzler *et al.*, 2002). However, there remains intrigue in the ClC field as further biophysical data revealed that certain members of this protein family, including the bacterial proteins used for these structural studies, function as chloride–proton exchangers rather than strict chloride channels (Accardi and Miller, 2004). To date, most evidence indicates that ClC-1 functions primarily as a true chloride channel (Picollo and Pusch, 2005).

B. Spectrum of *CLCN1* mutations in myotonia congenita

Figure 2.2 illustrates the location and type of most known *CLCN1* mutations (>120; Table 2.2) associated with myotonia congenita. Thomsen Disease is caused most often by heterozygous missense mutations. Notable exceptions are nonsense mutations such as R894X (George *et al.*, 1994) and E193X (Wu *et al.*, 2002), or certain frameshift mutations (de Diego *et al.*, 1999; Kuo *et al.*, 2006). By contrast, RGM is associated with the full spectrum of mutation types including missense, nonsense, and frameshifts. Mutations associated with either Thomsen Disease or RGM exhibit no topographical preferences in the channel protein.

Among the reported mutations associated with Thomsen Disease in North America, the nonconservative substitution of glycine-230 with glutamic acid (G230E) is most common (George *et al.*, 1993; Koty *et al.*, 1996). Historically interesting is the missense mutation P480L as it occurs in the extended pedigree of Julius Thomsen who originally described the syndrome (Steinmeyer *et al.*, 1994). Perhaps, the two most frequently discovered alleles in patients with

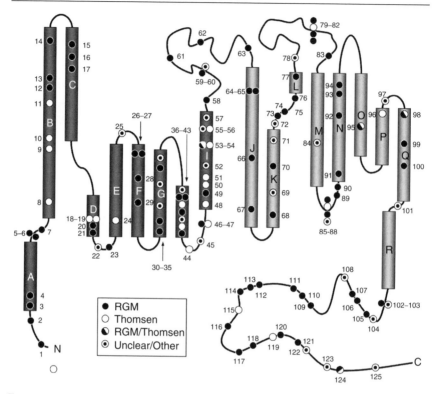

Figure 2.2. Location and type of mutations in the ClC-1 chloride channel associated with dominant myotonia congenita (Thomsen) and recessive generalized myotonia (RGM) illustrated on a transmembrane topology model of ClC-1. Each symbol represents the approximate position and type of a single mutation within the channel protein. Mutations where the inheritance is unclear or where the associated phenotype is atypical are denoted as "Unclear/Other." The numbering correlates with that of Table 2.2.

RGM are the missense mutation F413C and a 14-bp deletion producing a frameshift (Koch *et al.*, 1992; Meyer-Kleine *et al.*, 1994). Clustering of mutations in exon 8 suggests that this region is a potential hot spot for dominant mutations (Fialho *et al.*, 2007).

Most myotonia mutations are unique to individual families or isolated cases. Inheritance of a different mutant allele from each parent (*compound heterozygosity*) is common among subjects with RGM (Papponen *et al.*, 1999). In exceptional cases, families have been identified in which more than the two mutations segregate with the disease. This results in the occurrence of different intrafamilial compound genotypes that may in part explain the variability in myotonia severity within a single family (Sloan-Brown and George, 1997).

Table 2.2. CLCN1 Mutations Associated with Myotonia Congenita

No.	AA level	NA level	Exon	Topology	Phenotype	References
1	Altered splicing?	180 + 3, A to T	Intronic	N-terminus	RGM	Sloan-Brown and George (1997)
2	E67X	G199T	2	N-terminus	RGM	Fialho et al. (2007)
3	Q68X	C202T	2	N-terminus	RGM	Zhang et al. (1996)
4	Q74X	C220T	2	N-terminus	RGM	Mailänder et al. (1996)
5	Altered splicing?	302-1, G to A	Intronic	N-terminus	RGM	Trip et al. (2008)
6	Altered splicing?	302-2, A to C	Intronic	N-terminus	RGM	Trip et al. (2008)
7	R105C	C313T	3	N-terminus	RGM	Meyer-Kleine et al. (1995)
8	W118G	T352G	3	B helix	Thomsen	Lehmann-Horn et al. (1995)
9	M128V	A382G	3	B helix	Thomsen	Colding-Jorgensen et al. (2003)
10	A129T	G385A	3	B helix	Thomsen	Trip et al. (2008)
11	S132C	A394T	3	B helix	Thomsen	Wu et al. (2002)
12	D136G	A407G	3	B helix	RGM	Heine et al. (1994)
13	Y137X	C411G	3	B helix	RGM	Trip et al. (2008)
14	Y150C	A449G	4	B helix	RGM	Mailänder et al. (1996)
15	F161V	T481G	4	C helix	RGM	Plassart-Schiess et al. (1998)
16	V165G	T494G	4	C helix	RGM	Meyer-Kleine et al. (1995)
17	F167L	C501G	4	C helix	RGM	George et al. (1994)
18	E193X	G577T	5	D helix	Thomsen	Wu et al. (2002)
19	E193K	G577A	5	D helix	Thomsen	Colding-Jorgensen et al. (2003)
20	M194fs256X	585–589del	5	D helix	RGM	Trip et al. (2008)
21	L198V	C592G	5	D helix	RGM	Simpson et al. (2004)
22	G200R	G598A	5	D–E linker	Thomsen, fluctuating	Mailänder et al. (1996); Wagner et al. (1998)
23	G200fs258X	601/602insG	5	D–E linker	RGM	Esteban et al. (1998)
24	A218T	G652A	5	E helix	Thomsen	de Diego et al. (1999)
25	G230E	G689A	5	E–F linker	Thomsen, fluctuating	George et al. (1993; Lacomis et al. (1999)
26	Altered splicing?	696 + 2, T to A	Intronic	E helix	RGM	Brugnoni et al. (1999)
27	Altered splicing?	696 + 1, G to A	Intronic	E helix	RGM	McKay et al. (2006)

(Continues)

Table 2.2. (*Continued*)

No.	AA level	NA level	Exon	Topology	Phenotype	References
28	V236L	G706C	6	F helix	RGM	Kubisch et al. (1998)
29	C242X	T726A	6	F helix	RGM	Fialho et al. (2007)
30	Y261C	A782G	7	G helix	RGM	Mailänder et al. (1996)
31	Y263fs267X	789del	7	G helix	RGM	Trip et al. (2008)
32	T268M	C803T	7	G helix	Spontaneous, RGM?	Brugnoni et al. (1999)
33	C271R	T811C	7	G helix	Spontaneous	Fialho et al. (2007)
34	G276S	G826A	7	G helix	RGM	Fialho et al. (2007)
35	C277fs289X	831/832insG	7	G helix	"dystrophic" RGM	Nagamitsu et al. (2000)
36	L283F	C847T	7	H helix	Unclear	Wu et al. (2002)
37	Altered splicing?	G854A	8	H helix	RGM	Plassart-Schiess et al. (1998)
38	G285E	G854A	8	H helix	RGM	Kubisch et al. (1998) [reported as a splice site mutation by Plassart-Schiess et al. (1998)]
39	V286A	T857C	8	H helix	Thomsen	Kubisch et al. (1998)
40	F288S	T863C	8	H helix	RGM	Sun et al. (2001)
41	S289N	G866A	8	H helix	Spontaneous	Fialho et al. (2007)
42	I290M	C870G	8	H helix	Thomsen	Lehmann-Horn et al. (1995)
43	E291K	G871A	8	H helix	RGM	Meyer-Kleine et al. (1995)
44	F297S	T890C	8	H–I linker	Thomsen	Fialho et al. (2007)
45	V299L	G895C	8	H–I linker	RGM?	Fialho et al. (2007)
46	R300Q	G899A	8	H–I linker	Thomsen	Steinmeyer et al. (1994)
47	R300W	C898T	8	H–I linker	RGM	George et al. (1994)
48	W303R	T907C	8	I helix	Thomsen	Fialho et al. (2007)
49	G305E	G914A	8	I helix	RGM	Trip et al. (2008)
50	F306L	T916C	8	I helix	Thomsen	Fialho et al. (2007)
51	F307S	T920C	8	I helix	Thomsen	Kubisch et al. (1998)
52	T310M	C929T	8	I helix	Fluctuating myotonia	Wu et al. (2002)
53	A313V	C938T	8	I helix	Thomsen	Fialho et al. (2007)
54	A313T	G937A	8	I helix	RGM/Thomsen	Plassart-Schiess et al. (1998)
55	R317L	G950T	8	I helix	Spontaneous	Fialho et al. (2007)

(Continues)

Table 2.2. (Continued)

No.	AA level	NA level	Exon	Topology	Phenotype	References
56	R317Q	G950A	8	I helix	Thomsen	Meyer-Kleine et al. (1995)
57	A320V	C959T	8	I helix	Spontaneous	Fialho et al. (2007)
58	V321L	G961C	8	I–J linker	RGM	Fialho et al. (2007)
59	V327I	G979A	9	I–J linker	RGM	Lorenz et al. (1994)
60	Altered splicing?	979 +1, G to A	Intronic	I–J linker	Spontaneous, RGM?	Fialho et al. (2007)
61	I329T	T986C	9	I–J linker	RGM	Meyer-Kleine et al. (1995)
62	A331T	G991A	9	I–J linker	RGM	Sun et al. (2001)
63	R338Q	G1013A	9	I–J linker	RGM	George et al. (1994)
64	G355R	G1063? A or C	10	J helix	RGM	Deymeer et al. (1998)
65	Altered splicing?	1065-2, A to G	Intronic	J helix	RGM	Trip et al. (2008)
66	F365fs387X	1095–1096del 1096–1097del 1098–1099del	10	J helix	RGM	Meyer-Kleine et al. (1995)
67	R377X	C1129T	10	J helix	RGM	Fialho et al. (2007)
68	Altered splicing?	1167-10, T to C	Intronic	K helix	RGM	Trip et al. (2008)
69	P394fs427X	1183–1187del	11	K helix	Spontaneous	Fialho et al. (2007)
70	T398I	C1193T	11	K helix	RGM	Fialho et al. (2007)
71	A402V	C1205T	11	K helix	Spontaneous, Thomsen?	Fialho et al. (2007)
72	P408A	C1222G	11	K–L linker	Spontaneous	Fialho et al. (2007)
73	Q412P	A1235C	11	K–L linker	RGM	Morales et al. (2008)
74	F413C	T1238G	11	K–L linker	RGM	Koch et al. (1992)
75	A415V	C1244T	11	K–L linker	RGM	Mailänder et al. (1996)
76	E417G	A1250G	11	K–L linker	RGM	Trip et al. (2008)
77	P420fs429X	1262insC	12	L helix	RGM	Meyer-Kleine et al. (1995)
78	I424M	C1272G	12	L–M linker	Spontaneous, RGM?	Brugnoni et al. (1999)
79	F428S	T1283C	12	L–M linker	Thomsen	Wu et al. (2002)

(Continues)

Table 2.2. (*Continued*)

No.	AA level	NA level	Exon	Topology	Phenotype	References
80	L427fs433X	1278–1281del	12	L–M linker	RGM	Heine et al. (1994)
81	L427fs488X	1280–1283del	12	L–M linker	RGM	Heine et al. (1994)
82	L427fs465X	1279–1282del	12	L–M linker	RGM	Heine et al. (1994)
83	Q445X	C1333T	12	L–M linker	RGM	Sasaki et al. (1999)
84	S471F	C1412T	13	M helix	Thomsen?	Jou et al. (2004)
85	P480T	C1438A	13	M–N linker	Thomsen	Sasaki et al. (2001)
86	P480H	C1439A	13	M–N linker	Spontaneous	Fialho (2007)
87	P480L	C1439T	13	M–N linker	Thomsen	Steinmeyer et al. (1994)
88	I479fs503X	1437–1450del	13	M–N linker	RGM	Meyer-Kleine et al. (1994)
89	C481X	C1443A	13	M–N linker	RGM	Sangiuolo et al. (1998)
90	G482R	G1444A	13	M–N linker	RGM	Meyer-Kleine et al. (1995)
91	M485V	A1453G	13	N helix	RGM	Meyer-Kleine et al. (1995)
92	Altered splicing?	1471 + 1, G to A	Intronic	N helix	RGM	Meyer-Kleine et al. (1995)
93	R496S	G1488T	14	N helix	RGM	Lorenz et al. (1994)
94	G499R	G1495A	14	N helix	RGM	Zhang et al. (2000)
95	A531V	C1592T	15	O helix	RGM/Thomsen	Papponen et al. (1999)
96	T550M	C1649T	15	P helix	Thomsen	Sejersen et al. (1996)
97	Q552R	A1655G	15	P–Q linker	RGM/	Lehmann-Horn et al. (1995)
					Thomsen, myotonia levior	
98	I556N	T1667A	15	Q helix	RGM/Thomsen	Plassart-Schiess et al. (1998)
99	V563I	G1687A	15	Q helix	RGM	Sangiuolo et al. (1998)
100	A566T	G1696A	15	Q helix	RGM	Fialho et al. (2007)
101	P575S	C1723T	15	Q–R linker	Late-onset RGM?	Jou et al. (2004)
102	K614M	A1841T	16	C-terminus	RGM	Trip et al. (2008)
103	K614N	G1842C	16	C-terminus	RGM	Colding-Jorgensen et al. (2003)
104	D644G	A1931G	17	C-terminus	RGM?	Jou et al. (2004)
105	M646I	G1938A	17	C-terminus	RGM	Trip et al. (2008)

(*Continues*)

Table 2.2. (*Continued*)

No.	AA level	NA level	Exon	Topology	Phenotype	References
106	Q658X	C1972T	17	C-terminus	RGM	de Diego et al. (1999)
107	A659V	C1976T	17	C-terminus	RGM	Sasaki et al. (1999)
108	R669C	C2005T	17	C-terminus	Spontaneous, RGM?	de Diego et al. (1999)
109	F708L	C2124G	17	C-terminus	RGM	Sangiuolo et al. (1998)
110	D716fs793X	2149del	17	C-terminus	RGM	Plassart-Schiess et al. (1998)
111	Altered splicing?	2172 + 1, G to T	Intronic	C-terminus	RGM	Chen et al. (2004)
112	K752R	A2255G	18	C-terminus	RGM	Simpson et al. (2004)
113	Q754fs793X	2264del	18	C-terminus	RGM	Sangiuolo et al. (1998)
114	Altered splicing?	2284 + 5, C to T	Intronic	C-terminus	RGM	Sun et al. (2001)
115	L776fs793X	2330del	19	C-terminus	Thomsen	Kuo et al. (2006)
116	Q807X	C2419T	21	C-terminus	RGM	Chen and Chen (2001)
117	Altered splicing?	2452 + 2, T to A	Intronic	C-terminus	RGM	Sangiuolo et al. (1998)
118	C819X	C2457A	21	C-terminus	RGM	Trip et al. (2008)
119	H838fs872X	2513/2514insC	22	C-terminus	Thomsen	de Diego et al. (1999)
120	fs1870X	2518–2519del	23	C-terminus	RGM	Colding-Jorgensen et al. (2003)
121	G859D	G2576A	22	C-terminus	RGM	Deymeer et al. (1998)
122	Altered splicing?	2596-1, G to A	Intronic	C-terminus	RGM/Thomsen	Fialho et al. (2007)
123	P883T	C2647A	23	C-terminus	RGM?	Fialho et al. (2007)
124	R894X	C2680T	23	C-terminus	RGM/Thomsen	George et al. (1994)
125	P932L	C2795T	23	C-terminus	"dystrophic" RGM	Nagamitsu et al. (2000)

Single amino acid letter code with the original residue at the beginning, followed by the open reading frame position and the residue present in the mutant. Exon–intron boundaries are based on NM_000083 (GenBank). The secondary structure is adopted from Dutzler et al. *Abbreviations*: AA, amino acid; del, deletion of the residue(s); NA, nucleic acid; RGM, recessive generalized myotonia (Becker); X, termination codon; F365fs387X (example), frameshift at phenylalanine 365 producing nonsense residues until ORF position 387, where the protein is prematurely terminated; 601/602insG (example), guanine insertion between bases 601 and 602; 180 + 3, A to T (example), intronic mutation +3 downstream of the previous splice junction (upstream/following splice junction, if negative).

There are few reliable estimates of the carrier frequency of *CLCN1* mutations, but myotonia congenita is generally considered uncommon. Becker (1977) calculated a prevalence of approximately 1:23,000 and 1:50,000 for dominant and recessive myotonia congenita, respectively.

The allelic heterogeneity associated with myotonia congenita creates difficulty for performing molecular genetic analyses because a comprehensive survey of the entire *CLCN1* open reading frame, or possibly the full gene sequence, is necessary to detect mutations in an individual. For large or consanguineous families with dominant or recessive myotonia, linkage analysis with polymorphic markers in and around *CLCN1* on 7q35 may help establish a genetic diagnosis. However, most individuals with myotonia congenita appear sporadic or are associated with small kindreds, thus limiting the value of this approach.

A few mutations have been associated with either dominant or recessive myotonia congenita. Notable examples include the nonsense mutation R894X and several missense mutations (G200R, T268M, A313T, A531V, Q552R, and I556N). One explanation for this observation is incomplete penetrance or variable expressivity of certain dominant mutations. Variable phenotype severity and incomplete penetrance associated with certain dominant alleles have not been adequately explained. Current hypotheses include intrinsic differences in the severity of chloride channel dysfunction associated with specific mutations (Kubisch *et al.*, 1998) or the existence of modifier genes that affect chloride channel function or allelic expression (Duno *et al.*, 2004).

IV. PHYSIOLOGICAL BASIS OF MYOTONIA CONGENITA

Excitation and contraction of muscle are well understood at the cellular and molecular level. Rapidly opening voltage-gated sodium channels are responsible for the initial upstroke in the action potential, and somewhat slower activating potassium channels repolarize the membrane, thus bringing the cell to rest. Protection against spurious action potential triggering is provided by chloride channels that stabilize the membrane potential at the resting level. Myocytes must propagate action potentials deep into the muscle fiber to ensure near-simultaneous stimulation of the cell as a whole, a prerequisite for uniform contraction. Invaginations of the sarcolemma called *transverse* or T-tubules run radially into the myocyte, where they couple to the sarcoplasmic reticulum to trigger intracellular calcium release for initiating contraction. Although the T-tubular system is an ingenious invention of nature, the excitatory speed that it provides comes at a price. Due to the spatial confinements of the T-tubules, potassium exiting the cell during action potentials can accumulate extracellularly to the point of depolarizing the potassium equilibrium potential. Under normal physiological circumstances, the membrane potential is preserved owing to the counterbalancing effect of a high sarcolemmal chloride conductance. However, this countermeasure is crippled in myotonia and an elevated T-tubular potassium concentration promotes sarcolemmal hyperexcitability.

A. Role of the sarcolemmal chloride conductance

Experiments with myotonic goats (Adrian and Bryant, 1974) and computer simulations of muscle action potentials (Barchi, 1975) deduced that an 80% reduction in sarcolemmal chloride conductance will produce myotonia. The link between reduced chloride conductance and myotonia was further supported by the observation that normal muscle fibers became myotonic when bathed in an extracellular solution lacking chloride ions (Adrian and Bryant, 1974). Skeletal muscle has a high chloride conductance at rest accounting for approximately 70% of the total membrane ion conductance (Palade and Barchi, 1977). Chloride ions distribute passively across the sarcolemma such that chloride moves into or out of the cell until the intracellular concentration is adjusted to set the chloride equilibrium potential equal to the resting potential. Chloride ions do not actively set the resting potential. However, in response to any change in the membrane potential, a large chloride current will flow that tends to return the membrane potential to its previously established resting value. In skeletal muscle, the high resting chloride conductance acts as an electrical buffer that stabilizes the resting potential and will also promote repolarization after an action potential.

A reduced chloride conductance has two effects on sarcolemmal excitability. First, a lower electrical stimulus is required to elicit action potentials resulting in an enhanced excitability (Adrian and Bryant, 1974). Second, and more importantly in the context of myotonia, action potential propagation along the T-tubules brings with it an outwardly directed potassium flow that repolarizes the membrane. Due to the spatial confinement of the T-tubules, the potassium efflux elevates the local extracellular potassium concentration. The lowered chemical gradient (i.e., decreased driving force for potassium to leave the cell) reduces the efflux of positive charge, resulting in net depolarization and failure to reach the previously attained resting potential. This is referred to as *afterdepolarization* and it usually dissipates over several hundred milliseconds (Fig. 2.3). When sufficient numbers of impulses are propagated rapidly in a myotonic muscle fiber, the T-tubular potassium rises substantially and exerts a strong depolarizing effect on the membrane potential. Under normal conditions, this rise in extracellular potassium has little effect on the membrane voltage because of the electrical dampening effects of the sarcolemmal chloride conductance. However, in myotonic muscle where chloride conductance as a counterbalancing force has been diminished or abolished, the equilibrium potential for potassium accumulation in the T-tubular system becomes the sole determinant of the resting membrane potential. Under these conditions, the threshold voltage can be reached and

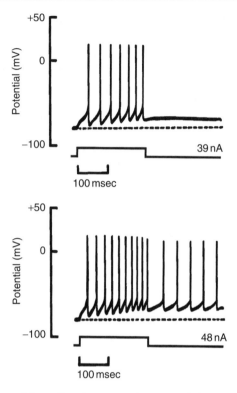

Figure 2.3. Action potential recordings from myotonic goat muscle in response to graded electrical stimulation (modified after Adrian and Bryant, 1974). In the top tracing, stimulation of muscle fibers (39 nA) produces multiple action potential spikes and a small afterdepolarization visible as the elevated membrane potential above the horizontal dotted line (representing the resting potential). With a stronger stimulus (48 nA, lower tracing), the afterdepolarization reaches threshold voltage and triggers spontaneous action potentials that fire after termination of the electrical stimulus.

spontaneous action potentials may be triggered in absence of neuromuscular transmission. These autonomous action potentials will cause persistent muscle contraction and delayed relaxation after a voluntary movement and result in clinical myotonia.

B. Functional consequences of *CLCN1* mutations

Experiments with recombinant human ClC-1 chloride channels heterologously expressed in *Xenopus oocytes* or cultured mammalian cells coupled with two-electrode voltage clamping or whole-cell patch clamping have helped ascertain the functional consequences of ClC-1 mutations. Recombinant human ClC-1 exhibits many of the physiological and biophysical properties of native sarcolemmal chloride channels when reconstituted in heterologous expression systems (Fahlke et al., 1996; Pusch et al., 1994). The channel has an ionic selectivity sequence of $Cl^- > Br^- > I^-$, and a current–voltage relationship exhibiting inward rectification (Fahlke et al., 1997b; Rychkov et al., 1998).

Approximately, half of the CLCN1 mutations associated with RGM are truncating alleles that have been generally assumed to produce nonfunctional channels. Certain mutations corrupt the cell surface delivery of mutant ClC-1 from the endoplasmic reticulum and this phenomenon explains the complete absence of measurable current in these mutants (Papponen et al., 2008).

Among the various missense mutations that have been characterized, most exhibit complete or near complete loss of function or a range of altered gating behaviors. One common gating defect caused by several recessive and dominant alleles is a shift in the voltage dependence of channel activation (Pusch et al., 1995). At rest (-85 mV in skeletal muscle), ClC-1 has a low (0.3) open probability. Upon depolarization, the likelihood of channels residing in an active or open state increases sigmoidally, reaching a maximum at positive potentials. Several mutations associated with either Thomsen Disease or RGM shift this relationship between membrane voltage and open probability to more depolarized potentials such that the activity of the mutant chloride channels remains low across the physiological range of membrane potentials (Pusch et al., 1995). The net effect of this altered voltage dependence is a physiological loss of function.

Another more unusual type of gating disturbance observed for recessive ClC-1 mutations is an inversion of the voltage sensitivity of channel activation. This was first described for the recessive mutation D136G (Fahlke et al., 1995), but has since been reported with other alleles (Zhang et al., 2000). Finally, a distinct mechanism of chloride channel dysfunction associated with the common dominant allele G230E has been elucidated (Fahlke et al., 1997a). This mutation causes a dramatic alteration in the ion selectivity of the mutant chloride channel, resulting in a reversal in the normal halide ion selectivity sequence and an increase in cation permeability. Disruption of ion selectivity in this context likely originates from a major structural disruption in segments near the pore of the channel (Fahlke et al., 1997d). The G230E mutation introduces a novel negative charge in close proximity to a glutamate residue considered to be a critical component of the fast gate (Dutzler et al., 2003).

In heterozygous mutation carriers, both wildtype and mutant chloride channels coexist in the same cell. Experiments replicating the combination of wildtype and mutant alleles in vitro have illustrated that the presence of a single wildtype allele can partially compensate for the defect associated with recessive mutations. By contrast, if the rogue allele interferes with the functionality of its healthy partner, then dominant inheritance ensues (Thomsen Disease) as the chloride conductance is suppressed close to the critical threshold (Kubisch et al., 1998). This apparent interaction between alleles most likely occurs at the protein level.

The dimeric architecture of ClC-1 helps explain the dominant-negative reduction in chloride channel activity seen for Thomsen Disease mutations. Assuming that mutant/wildtype dimers are inactive and that wildtype (WT) and mutant (MUT) ClC-1 subunits randomly assort and assemble, heterozygosity will result in a 75% reduction in chloride conductance in a 1:2:1 pool of WT/WT:MUT/WT:MUT/MUT channels. This comes close to mathematically predicted 80% reduction needed to promote myotonia. Recessive alleles as well as certain dominant alleles that exhibit incomplete penetrance in Thomsen Disease families (Kubisch et al., 1998) can be partially rescued by assembly with the wildtype allele. This has led to the hypothesis that the biophysical mechanism of incomplete penetrance may relate to a milder degree of dominant-negative effects of the mutant allele on the dimeric channel complex.

C. Unanswered questions

Major questions relevant to the pathophysiology of myotonia congenita remain unanswered. For example, explanations for variable expressivity and incomplete penetrance are currently insufficient as in most Mendelian disorders. The leading hypotheses to explain these phenomena in myotonia congenita include intrinsic variability of channel dysfunction and the variable impact that mutant alleles have on the wildtype channel protein. Variation in allelic expression has also been proposed (Duno et al., 2004).

Another important but unsolved issue is the etiology of the warm-up phenomenon. There have been suggestions that an activity-dependent increase in ClC-1 chloride channel activity may underlie the warm-up phenomenon in some cases (Pusch et al., 1995) although this is difficult to reconcile with myotonia caused by a complete absence of chloride channel protein. It is conceivable that other types of chloride channels contribute to the sarcolemmal chloride conductance in an activity-dependent fashion, although there is little evidence to support this presently. Another hypothesis is that the same potassium accumulation that causes afterdepolarization may simultaneously inactivate a fraction of sodium channels. Once inactivated, sodium channels need hyperpolarization to reprime for activity such that the pool of sodium channels available for opening is reduced. Functionally, this would render the muscle less excitable despite elevated T-tubular potassium.

Another plausible physiological basis for warm-up is activation of the Na^+/K^+-ATPase (sodium pump) during repetitive action potential generation. The thought here is that the cell compensates for afterdepolarizations by increasing potassium uptake in the T-tubules. Because the sodium pump is electrogenic (e.g., transports two K^+ ions for every three Na^+ ions), it has a repolarizing effect on the membrane potential. Although this mechanism seems plausible, *in vivo* exposure of myotonic muscle to an inhibitor of Na^+/K^+-ATPase (ouabain) does not alter the warm-up effect (Van Beekvelt *et al.*, 2006).

V. MYOTONIA CONGENITA IN ANIMAL MODELS

Our current understanding of the pathogenesis of myotonia caused by defects in chloride channel function originated with investigations by Brown and Harvey (1939), who were the first to make electromyographic recordings from myotonic goats. These animals have been variably referred to as "fainting," "nervous," "stiff-legged," or "epileptic" goats because of their tendency to develop severe acute muscular stiffness followed by loss of balance and falling when attempting to make sudden forceful movements or when startled (White and Plaskett, 1904). The term "fainting" should not be taken literally as the animals show no tendency to lose consciousness. Anecdotal evidence suggests that the myotonic goat phenotype is transmitted as an autosomal-dominant trait although many herds have been inbred resulting in behaviors reminiscent of human RGM (Kolb, 1938). Also, like in the human disease, myotonia congenita in goats is caused by a missense mutation in the muscle chloride channel gene (Beck *et al.*, 1996). Specifically, a proline substitution for a highly conserved alanine residue in the carboxyl terminus of goat ClC-1 gives rise to a large (+45 mV) depolarizing shift in activation similar to several human myotonia mutations. Coexpression of the myotonic goat mutation with the wildtype channel was not performed leaving unsettled whether this mutant allele exhibits a recessive or dominant character *in vitro*. Preliminary evidence indicated that this mutation was the only allele among myotonic goats in Tennessee (Beck *et al.*, 1996) where the animals were first described (White and Plaskett, 1904).

In the 1960s and 1970s, Bryant, Lipicky, and colleagues determined the membrane abnormalities in myotonic goats. They discovered that the resting membrane chloride conductance in skeletal muscle fibers was severely decreased and this was directly associated with myotonic discharges (Adrian and Bryant, 1974; Lipicky and Bryant, 1966). The link between reduced chloride conductance and myotonia was strengthened by their observation that normal muscle fibers became myotonic when bathed in an extracellular solution lacking chloride ions. The role of the T-tubule was also inferred from experiments performed on detubulated myotonic muscle. Further support for the chloride hypothesis came from the finding that the myotonia-inducing herbicide 2,4-D suppressed sarcolemmal chloride conductance (Furman and Barchi, 1978). Shortly thereafter, Bretag *et al.* (1980) demonstrated that muscle chloride channel block also led to a myotonic response in frogs.

By the end of the 1970s, various mouse models of congenital myotonia had become available. The first murine myotonia line arose spontaneously and became known as "arrested development of righting response" or *adr* (Watts *et al.*, 1978). Several similar models followed, including another spontaneous line, *adr^mto* (Heller *et al.*, 1982), the ethylnitrosourea mutagenesis-derived *adr^K* line (Neumann and Weber, 1989), and *adr^crp*, which appears in the literature only as a personal communication with little additional information (Goblet and Whalen, 1995). Owing to tight phenotypic similarities, the models were predicted and confirmed to be allelic (Jockusch *et al.*, 1988), and later linked to mutations in the mouse skeletal muscle chloride channel gene. Specifically, *adr* mice have a transposon insertion that disrupts the coding region of murine ClC-1 (Steinmeyer *et al.*, 1991a), whereas *adr^mto* is caused by a nonsense mutation (R47X) (Gronemeier *et al.*, 1994) and *adr^K* is associated with a missense allele, I553T (Gronemeier *et al.*, 1994). Mutation data on *adr^crp* have not been reported.

Chloride channel mutations have also been discovered in two breeds of dogs (miniature Schnauzer and Australian cattle dog) with recessive myotonia congenita (Finnigan *et al.*, 2007; Rhodes *et al.*, 1999). A missense mutation in canine ClC-1 associated with myotonia in Schnauzers was demonstrated to exhibit a large shift in the voltage dependence of activation with substantial rescue during coexpression with the wildtype allele as observed with various human RGM mutation (Rhodes *et al.*, 1999). Canine myotonia congenita could provide an excellent model for testing pharmacological and other novel therapeutic approaches such as gene therapy (Rogers *et al.*, 2002).

VI. TREATMENT OF MYOTONIA CONGENITA

Many individuals with myotonia do not require any pharmacological intervention. Most patients prefer to minimize their symptoms by avoiding situations and stimuli that trigger myotonic episodes. When these maneuvers are insufficient, drugs that reduce the excitability of the sarcolemma can be used.

A. Pharmacological therapies

Historically, quinidine and quinine were used as antimyotonic agents. These drugs were staples in the treatment of cardiac arrhythmia (Brodsky *et al.*, 1996) and malaria (Brossi *et al.*, 1971), and usually well tolerated in low-dose, short-term applications. However, continued administration of these agents to subdue the myotonic response is not recommended owing to *cinchonism*, a well-described toxic reaction that includes visual and acoustic impairments, vertigo, and various gastrointestinal symptoms. Prolonged use of quinine/quinidine is furthermore ototoxic (Brown and Feldman, 1978) and may also lead to severe neurological manifestations (Gilbert, 1978; Johnson et al., 1990) or even death (Thomson, 1956). This leaves the option of employing these compounds temporarily as needed, for example in anticipation of situations where uninhibited motor performance is essential. Unfortunately, the antimyotonic effects of quinidine and quinine wane with repeated use such that the long-term benefit in treating myotonia is limited.

Other drugs that have been used in treating myotonia with varying levels of success include procaine (Munsat, 1967), tocainide (Catalano et al., 2008; Rüdel et al., 1980), mexiletine (Ricker et al., 1994), carbamazepine (Savitha et al., 2006), and phenytoin (Aichele et al., 1985). All of these drugs act by use-dependent block of voltage-gated sodium channels. Mexiletine, starting at 200 mg three times daily, is the preferred drug (Ricker et al., 1994), but controlled trials comparing various agents have not been performed. There is also some work on the benefits of dehydroepiandrosterone, which may also block sodium currents while minimizing associated loss in strength (Nakazora and Kurihara, 2005). As with all compounds that modulate sodium currents, special consideration has to be given to elderly patients or those with known cardiac conduction abnormalities. In all cases, close neurological and cardiac monitoring of the patient ought to be in place when the therapy is started and at routine intervals to minimize the risks from unwanted drug actions.

A directed approach for ameliorating myotonia due to chloride channel defects would be to pharmacologically increase the chloride conductance of skeletal muscle. Taurine and the R-$(+)$ isomer of clofibric acid produce measurable increases in the resting chloride conductance of skeletal muscle. The effect is modest and not sufficient to prevent myotonia (Conte-Camerino et al., 1984), although prior data suggested improvement with some of these compounds (Conte Camerino et al., 1989).

There is a need for a controlled drug trial to assess the efficacy and tolerability of available antimyotonic drugs (Trip et al., 2006). The fact that no such study has thus far been conducted is in part a reflection of the rarity of myotonia congenita. Further, it is exceedingly complicated to establish a method that reliably assesses the myotonic response as well as to control for wide intersubject variation and temporal fluctuations in symptom severity.

B. Other therapeutic considerations

Nonpharmacological therapies may be very beneficial in the treatment of myotonia congenita. The emotional state of affected individuals may influence phenotype severity with stress and mental discomfort contributing to worsen myotonia. Relaxation techniques may be of benefit as long as there is no attempt to imply that the etiology of myotonia is psychological. The benefits of mental relaxation may explain improvements observed with alcohol use in some subjects (Becker, 1977). Exercises that improve flexibility may also have benefits, in particular avoiding muscle strains, when movements are performed during myotonic events.

Anesthesia should be administered cautiously to subjects with myotonia congenita as there is an increased risk for a malignant hyperthermia-like syndrome (Becker, 1977; Farbu *et al.*, 2003; Heiman-Patterson *et al.*, 1988). Not surprisingly, the list of contraindicated compounds includes drugs that are known to worsen myotonia, such as propofol and gabapentin (Allford, 2007; Speedy, 1990). For general anesthesia in myotonic subjects, special consideration must be given to these risks.

C. Gene therapy for myotonia congenita

With increased knowledge regarding the molecular basis of myotonia congenita, novel treatment strategies aimed at correcting the molecular defect are conceivable. Conceptually, the treatment of a recessive disorder such as RGM may be feasible by introducing a functional copy of the normal gene into affected tissues. However, this approach may not be fully effective in disorders caused by a dominant-negative mechanism and other "gene repair" strategies have been proposed. In one study, Rogers *et al.* (2002) successfully repaired a ClC-1 missense mutation *in vitro* by employing a trans-splicing ribozyme that catalyzed the exchange of a wildtype RNA sequence to a targeted mutant mRNA. While the efficacy of the repair reaction was limited, this proof-of-principle study demonstrated that gene therapy for myotonia may be feasible. A major challenge to successful gene therapy of myotonia congenita remains the great difficulty in systemically targeting skeletal muscle with a gene delivery vector.

VII. CONCLUDING REMARKS

Myotonia congenita was the first recognized inherited channelopathy involving a ClC chloride channel. Unraveling the molecular basis for this disease has provided insight into the normal physiology of sarcolemmal excitability, and the structure–function relationships of the ClC-1 chloride channel. Current challenges exist in promoting improved recognition of myotonia congenita by health care providers, and exploiting the therapeutic opportunities that have arisen since the elucidation of the molecular genetic basis of this disorder.

Acknowledgments

We thank Isaac Sanchez, B. S., for his assistance with the mutation tables and for the artwork.

References

Abdalla, J. A., Casley, W. L., Cousin, H. K., Hudson, A. J., Murphy, E. G., Cornelis, F. C., Hashimoto, L., and Ebers, G. C. (1992). Linkage of Thomsen disease to the T-cell-receptor beta (TCRB) locus on chromosome 7q35. *Am. J. Hum. Genet.* **51,** 579–584.

Accardi, A., and Miller, C. (2004). Secondary active transport mediated by a prokaryotic homologue of ClC Cl⁻ channels. *Nature* **427,** 803–807.

Adrian, R. H., and Bryant, S. H. (1974). On the repetitive discharge in myotonic muscle fibres. *J. Physiol.* **240,** 505–515.

Aichele, R. Aichele, R., Paik, H., and Heller, A. H. (1985). Efficacy of phenytoin, procainamide, and tocainide in murine genetic myotonia. *Exp. Neurol.* **87,** 377–381.

Allford, M. A. (2007). Prolonged myotonia and dystonia after general anaesthesia in a patient taking gabapentin. *Br. J. Anaesth.* **99,** 218–220.

Barchi, R. L. (1975). Myotonia. An evaluation of the chloride hypothesis. *Arch. Neurol.* **32,** 175–180.

Beck, C. L., Fahlke, C., and George, A. L., Jr. (1996). Molecular basis for decreased muscle chloride conductance in the myotonic goat. *Proc. Natl Acad. Sci. USA* **93,** 11248–11252.

Becker, P. E. (1966). Zur Genetik der Myotonien. *In* "Progressive Muskeldystrophie–Myotonie–Myasthenie" (E. Kuhn, ed.), pp. 247–255. Springer Verlag, Berlin.

Becker, P. E. (1977). "Myotonia Congenita and Syndromes Associated with Myotonia." Thieme, Stuttgart.

Bradberry, S. M., Watt, B. E., Proudfoot, A. T., and Vale, J. A. (2000). Mechanisms of toxicity, clinical features, and management of acute chlorophenoxy herbicide poisoning: A review. *J. Toxicol. Clin. Toxicol.* **38,** 111–122.

Bretag, A. H., Dawe, S. R., and Moskwa, A. G. (1980). Chemically induced myotonia in amphibia. *Nature* **286,** 625–626.

Brodsky, M. A., Chun, J. G., Podrid, P. J., Douban, S., Allen, B. J., and Cygan, R. (1996). Regional attitudes of generalists, specialists, and subspecialists about management of atrial fibrillation. *Arch. Intern. Med.* **156,** 2553–2562.

Brossi, A., Uskokovic, M., Gutzwiller, J., Krettli, A. U., and Brener, Z. (1971). Antimalarial activity of natural, racemic and unnatural dihydroquinine, dihydroquinidine and their various racemic analogs in mice infected with *Plasmodium berghei. Experientia* **27,** 1100–1101.

Brown, R. D., and Feldman, A. M. (1978). Pharmacology of hearing and ototoxicity. *Annu. Rev. Pharmacol. Toxicol.* **18,** 233–252.

Brown, G. L., and Harvey, A. M. (1939). Congenital myotonia in the goat. *Brain* **62,** 341–363.

Brugnoni, R., Galantini, S., Confalonieri, P., Balestrini, M. R., Cornelio, F., and Mantegazza, R. (1999). Identification of three novel mutations in the major human skeletal muscle chloride channel gene (CLCN1), causing myotonia congenita. *Hum. Mutat.* **14,** 447.

Bryant, S. H. (1962). Muscle membrane of normal and myotonic goats in normal and low external chloride. *Fed. Proc.* **21,** 312.

Bryant, S. H., and Morales-Aguilera, A. (1971). Chloride conductance in normal and myotonic muscle fibres and the action of monocarboxylic aromatic acids. *J. Physiol.* **219,** 367–383.

Catalano, A., Carocci, A., Corbo, F., Franchini, C., Muraglia, M., Scilimati, A., De Bellis, M., De Luca, A., Camerino, D. C., Sinicropi, M. S., and Tortorella, V. (2008). Constrained analogues of tocainide as potent skeletal muscle sodium channel blockers towards the development of antimyotonic agents. *Eur. J. Med. Chem.* doi:10.1016/j.ejmech.2008.01.023.

Chang, T. Y., Kuo, H. C., Hsiao, K. M., and Huang, C. C. (2007). Phenotypic variability of autosomal dominant myotonia congenita in a Taiwanese family with muscle chloride channel (CLCN1) mutation. *Acta Neurol. Taiwan* **16,** 214–220.

Chen, M. F., and Chen, T. Y. (2001). Different fast-gate regulation by external Cl- and H+ of the muscle-type ClC chloride channels. *J. Gen. Physiol.* **118**, 23-32.

Chen, L., Schaerer, M., Lu, Z. H., Lang, D., Joncourt, F., Weis, J., Fritschi, J., Kappeler, L., Gallati, S., Sigel, E., and Burgunder, J. M. (2004). Exon 17 skipping in CLCN1 leads to recessive myotonia congenita. *Muscle Nerve* **29**, 670–676.

Colding-Jorgensen, E. (2005). Phenotypic variability in myotonia congenita. *Muscle Nerve* **32**, 19–34.

Colding-Jorgensen, E., Dun, O. M., Schwartz, M., and Vissing, J. (2003). Decrement of compound muscle action potential is related to mutation type in myotonia congenita. *Muscle Nerve* **27**, 449–455.

Conte-Camerino, D., Tortorella, V., Ferranini, E., and Bryant, S. H. (1984). The toxic effects of clofibrate and its metabolite on mammalian skeletal muscle: An electrophysiological study. *Arch. Toxicol. Suppl.* **7**, 482–484.

Conte Camerino, D., De Luca, A., Mambrini, M., Ferrannini, E., Franconi, F., Giotti, A., and Bryant, S. H. (1989). The effects of taurine on pharmacologically induced myotonia. *Muscle Nerve* **12**, 898–904.

de Diego, C., Gamez, J., Plassart-Schiess, E., Lasa, A., Del Rio, E., Cervera, C., Baiget, M., Gallano, P., and Fontaine, B. (1999). Novel mutations in the muscle chloride channel CLCN1 gene causing myotonia congenita in Spanish families. *J. Neurol.* **246**, 825–829.

Deymeer, F., Cakirkaya, S., Serdaroglu, P., Schleithoff, L., Lehmann-Horn, F., Rüdel, R., and Ozdemir, C. (1998). Transient weakness and compound muscle action potential decrement in myotonia congenita. *Muscle Nerve* **21**, 1334–1337.

Duno, M., Colding-Jorgensen, E., Grunnet, M., Jespersen, T., Vissing, J., and Schwartz, M. (2004). Difference in allelic expression of the CLCN1 gene and the possible influence on the myotonia congenita phenotype. *Eur. J. Hum. Genet.* **12**, 738–743.

Dutzler, R., Campbell, E. B., Cadene, M., Chait, B. T., and MacKinnon, R. (2002). X-ray structure of a ClC chloride channel at 3.0 A reveals the molecular basis of anion selectivity. *Nature* **415**, 287–294.

Dutzler, R., Campbell, E. B., and MacKinnon, R. (2003). Gating the selectivity filter in ClC chloride channels. *Science* **300**, 108–112.

Engel, W. K., and Brooke, M. H. (1966). Histochemistry of the myotonic disorders. *In* "Progressive Muskeldystrophie–Myotonie–Myasthenie" (E. Kuhn, ed.), pp. 203–222. Springer Verlag, Berlin.

Esteban, J., Neumeyer, A. M., McKenna-Yasek, D., and Brown, R. H. (1998). Identification of two mutations and a polymorphism in the chloride channel CLCN-1 in patients with Becker's generalized myotonia. *Neurogenetics* **1**, 185–188.

Fahlke, C., Rüdel, R., Mitrovic, N., Zhou, M., and George, A. L., Jr. (1995). An aspartic acid residue important for voltage-dependent gating of human muscle chloride channels. *Neuron* **15**, 463–472.

Fahlke, C., Rosenbohn, A., Mitrovic, N., George, A. L., Jr., and Rüdel, R. (1996). Mechanism of voltage-dependent gating in skeletal muscle chloride channels. *Biophys. J.* **71**, 695–706.

Fahlke, C., Beck, C. L., and George, A. L., Jr. (1997a). A mutation in autosomal dominant myotonia congenita affects pore properties of the muscle chloride channel. *Proc. Natl Acad. Sci. USA* **94**, 2729–2734.

Fahlke, C., Durr, C., and George, A. L., Jr. (1997b). Mechanism of ion permeation in skeletal muscle chloride channels. *J. Gen. Physiol.* **110**, 551–564.

Fahlke, C., Knittle, T., Gurnett, C. A., Campbell, K. P., and George, A. L., Jr. (1997c). Subunit stoichiometry of human muscle chloride channels. *J. Gen. Physiol.* **109**, 93–104.

Fahlke, C., Yu, H. T., Beck, C. L., Rhodes, T. H., and George, A. L., Jr. (1997d). Pore-forming segments in voltage-gated chloride channels. *Nature* **390**, 529–532.

Farbu, E., Softeland, E., and Bindoff, L. A. (2003). Anaesthetic complications associated with myotonia congenita: Case study and comparison with other myotonic disorders. *Acta Anaesthesiol. Scand.* **47**, 630–634.

Fialho, D., Schorge, S., Pucovska, U., Davies, N. P., Labrum, R., Haworth, A., Stanley, E., Sud, R., Wakeling, W., Davis, M. B., Kullmann, D. M., and Hanna, M. G. (2007). Chloride channel myotonia: Exon 8 hot-spot for dominant-negative interactions. *Brain* **130**, 3265–3274.

Fialho, D., Kullmann, D. M., Hanna, M. G., and Schorge, S. (2008). Non-genomic effects of sex hormones on CLC-1 may contribute to gender differences in myotonia congenita. *Neuromuscul. Disord.* **18**, 869–872.

Finnigan, D. F., Hanna, W. J., Poma, R., and Bendall, A. J. (2007). A novel mutation of the CLCN1 gene associated with myotonia hereditaria in an Australian cattle dog. *J. Vet. Intern. Med.* **21**, 458–463.

Freeble, C. R., Jr., and Rini, J. M. (1949). Myotonia congenita (Thomsen's disease) report of a case. *Ohio Med.* **45**, 459.

Furman, R. E., and Barchi, R. L. (1978). The pathophysiology of myotonia produced by aromatic carboxylic acids. *Ann. Neurol.* **4**, 357–365.

George, A. L., Jr., Crackower, M. A., Abdalla, J. A., Hudson, A. J., and Ebers, G. C. (1993). Molecular basis of Thomsen's disease (autosomal dominant myotonia congenita). *Nat. Genet.* **3**, 305–310.

George, A. L., Jr., Sloan-Brown, K., Fenichel, G. M., Mitchell, G. A., Spiegel, R., and Pascuzzi, R. M. (1994). Nonsense and missense mutations of the muscle chloride channel gene in patients with myotonia congenita. *Hum. Mol. Genet.* **3**, 2071–2072.

Gilbert, G. J. (1978). Quinidine dementia. *Am. J. Cardiol.* **41**, 791.

Goblet, C., and Whalen, R. G. (1995). Modifications of gene expression in myotonic murine skeletal muscle are associated with abnormal expression of myogenic regulatory factors. *Dev. Biol.* **170**, 262–273.

Gronemeier, M., Condie, A., Prosser, J., Steinmeyer, K., Jentsch, T. J., and Jockusch, H. (1994). Nonsense and missense mutations in the muscular chloride channel gene Clc-1 of myotonic mice. *J. Biol. Chem.* **269**, 5963–5967.

Gutmann, L., and Phillips, L. H., II (1991). Myotonia congenita. *Semin. Neurol.* **11**, 244–248.

Hanke, W., and Miller, C. (1983). Single chloride channels from *Torpedo* electroplax. Activation by protons. *J. Gen. Physiol.* **82**, 25–45.

Heiman-Patterson, T., Martino, C., Rosenberg, H., Fletcher, J., and Tahmoush, A. (1988). Malignant hyperthermia in myotonia congenita. *Neurology* **38**, 810–812.

Heller, A. H., Eicher, E. M., Hallett, M., and Sidman, R. L. (1982). Myotonia, a new inherited muscle disease in mice. *J. Neurosci.* **2**, 924–933.

Heine, R., George, A. L., Jr., Pika, U., Deymeer, F., Rüdel, R., and Lehmann-Horn, F. (1994). Proof of a non-functional muscle chloride channel in recessive myotonia congenita (Becker) by detection of a four base-pair deletion. *Hum. Mol. Genet.* **3**, 1123–1128.

Jentsch, T. J., Steinmeyer, K., and Schwarz, G. (1990). Primary structure of *Torpedo marmorata* chloride channel isolated by expression cloning in *Xenopus* oocytes. *Nature* **348**, 510–514.

Jockusch, H., Bertram, K., and Schenk, S. (1988). The genes for two neuromuscular diseases of the mouse, 'arrested development of righting response', adr, and 'myotonia', mto, are allelic. *Genet. Res.* **52**, 203–205.

Johnson, A. G., Day, R. O., and Seldon, W. A. (1990). A functional psychosis precipitated by quinidine. *Med. J. Aust.* **153**, 47–49.

Jou, S. B., Chang, L. I., Pan, H., Chen, P. R., and Hsiao, K. M. (2004). Novel CLCN1 mutations in Taiwanese patients with myotonia congenita. *J. Neurol.* **251**, 666–670.

Koch, M. C., Steinmeyer, K., Lorenz, C., Ricker, K., Wolf, F., Otto, M., Zoll, B., Lehmann-Horn, F., Grzeschik, K. H., and Jentsch, T. J. (1992). The skeletal muscle chloride channel in dominant and recessive human myotonia. *Science* **257,** 797–800.

Kolb, L. C. (1938). Congenital myotonia in goats. *Bull. Johns Hopkins Hosp.* **63,** 221–237.

Koty, P. P., Pegoraro, E., Hobson, G., Marks, H. G., Turel, A., Flagler, D., Cadaldini, M., Angelini, C., and Hoffman, E. P. (1996). Myotonia and the muscle chloride channel: Dominant mutations show variable penetrance and founder effect. *Neurology* **47,** 963–968.

Kubisch, C., Schmidt-Rose, T., Fontaine, B., Bretag, A. H., and Jentsch, T. J. (1998). ClC-1 chloride channel mutations in myotonia congenita: Variable penetrance of mutations shifting the voltage dependence. *Hum. Mol. Genet.* **7,** 1753–1760.

Kuo, H. C., Hsiao, K. M., Chang, L. I., You, T. H., Yeh, T. H., and Huang, C. C. (2006). Novel mutations at carboxyl terminus of ClC-1 channel in myotonia congenita. *Acta Neurol. Scand.* **113,** 342–346.

Kwiecinski, H. (1981). Myotonia induced by chemical agents. *Crit. Rev. Toxicol.* **8,** 279–310.

Lacomis, D., Gonzales, J. T., and Giuliani, M. J. (1999). Fluctuating clinical myotonia and weakness from Thomsen's disease occurring only during pregnancies. *Clin. Neurol. Neurosurg.* **101,** 133–136.

Lehmann-Horn, F., Mailander, V., Heine, R., and George, A. L. (1995). Myotonia levior is a chloride channel disorder. *Hum. Mol. Genet.* **4,** 1397–1402.

Lipicky, R. J., and Bryant, S. H. (1966). Sodium, potassium, and chloride fluxes in intercostal muscle from normal goats and goats with hereditary myotonia. *J. Gen. Physiol.* **50,** 89–111.

Lorenz, C., Meyer-Kleine, C., Steinmeyer, K., Koch, M. C., and Jentsch, T. J. (1994). Genomic organization of the human muscle chloride channel ClC-1 and analysis of novel mutations leading to Becker-type myotonia. *Hum. Mol. Genet.* **3,** 941–946.

Mailänder, V., Heine, R., Deymeer, F., and Lehmann-Horn, F. (1996). Novel muscle chloride channel mutations and their effects on heterozygous carriers. *Am. J. Hum. Genet.* **58,** 317–324.

Mankodi, A., Takahashi, M. P., Jiang, H., Beck, C. L., Bowers, W. J., Moxley, R. T., Cannon, S. C., and Thornton, C. A. (2002). Expanded CUG repeats trigger aberrant splicing of ClC-1 chloride channel pre-mRNA and hyperexcitability of skeletal muscle in myotonic dystrophy. *Mol. Cell* **10,** 35–44.

Mastaglia, F. L. (1982). Adverse effects of drugs on muscle. *Drugs* **24,** 304–321.

McKay, O. M., Krishnan, A. V., Davis, M., and Kiernan, M. C. (2006). Activity-induced weakness in recessive myotonia congenita with a novel (696+1G>A) mutation. *Clin. Neurophysiol.* **117,** 2064–2068.

Meyer-Kleine, C., Ricker, K., Otto, M., and Koch, M. C. (1994). A recurrent 14 bp deletion in the CLCN1 gene associated with generalized myotonia (Becker). *Hum. Mol. Genet.* **3,** 1015–1016.

Meyer-Kleine, C., Steinmeyer, K., Ricker, K., Jentsch, T. J., and Koch, M. C. (1995). Spectrum of mutations in the major human skeletal muscle chloride channel gene (CLCN1) leading to myotonia. *Am. J. Hum. Genet.* **57,** 1325–1334.

Miller, C. (1982). Open-state substructure of single chloride channels from *Torpedo* electroplax. *Philos. Trans. R. Soc. Lond. B Biol. Sci.* **299,** 401–411.

Miller, C., and White, M. M. (1984). Dimeric structure of single chloride channels from *Torpedo* electroplax. *Proc. Natl Acad. Sci. USA* **81,** 2772–2775.

Morales, F., Cuenca, P., del Valle, G., Vasques, M., Brian, R., Sittenfeld, M., Johnson, K., Lin, X., and Ashizawa, T. (2008). Gene symbol: CLCN1—Disease: Myotonia congenita. *Hum. Genet.* **123,** 104–105.

Morgan, K. G., Entrikin, R. K., and Bryant, S. H. (1975). Myotonia and block of chloride conductance by iodide in avian muscle. *Am. J. Physiol.* **229,** 1155–1158.

Munsat, T. L. (1967). Therapy for myotonia: A double-blind evaluation of diphenylhydantoin, procainimide, and placebo. *Neurology* **17,** 359–367.

Nagamitsu, S., Matsuura, T., Khajavi, M., Armstrong, R., Gooch, C., Harati, Y., and Ashizawa, T. (2000). A "dystrophic" variant of autosomal recessive myotonia congenita caused by novel mutations in the CLCN1 gene. *Neurology* **55,** 1697–1703.

Nakazora, H., and Kurihara, T. (2005). The effect of dehydroepiandrosterone sulfate (DHEAS) on myotonia: Intracellular studies. *Intern. Med.* **44,** 1247–1251.

Neumann, P., and Weber, T. (1989). A new allele of adr. *Mouse News Lett.* **83,** 157.

Nobile, M., Pusch, M., Rapisarda, C., and Ferroni, S. (2000). Single-channel analysis of a ClC-2-like chloride conductance in cultured rat cortical astrocytes. *FEBS Lett.* **479,** 10–14.

Palade, P. T., and Barchi, R. L. (1977). On the inhibition of muscle membrane chloride conductance by aromatic carboxylic acids. *J. Gen. Physiol.* **69,** 879–896.

Papponen, H., Toppinen, T., Baumann, P., Myllyla, V., Leisti, J., Kuivaniemi, H., Tromp, G., and Myllyla, R. (1999). Founder mutations and the high prevalence of myotonia congenita in northern Finland. *Neurology* **53,** 297–302.

Papponen, H., Nissinen, M., Kaisto, T., Myllyla, V. V., Myllyla, R., and Metsikko, K. (2008). F413C and A531V but not R894X myotonia congenita mutations cause defective endoplasmic reticulum export of the muscle-specific chloride channel CLC-1. *Muscle Nerve* **37,** 317–325.

Picollo, A., and Pusch, M. (2005). Chloride/proton antiporter activity of mammalian CLC proteins ClC-4 and ClC-5. *Nature* **436,** 420–423.

Pierno, S., De Luca, A., Tricarico, D., Roselli, A., Natuzzi, F., Ferrannini, E., Laico, M., and Camerino, D. C. (1995). Potential risk of myopathy by HMG-CoA reductase inhibitors: A comparison of pravastatin and simvastatin effects on membrane electrical properties of rat skeletal muscle fibers. *J. Pharmacol. Exp. Ther.* **275,** 1490–1496.

Pierno, S., Didonna, M. P., Cippone, V., De Luca, A., Pisoni, M., Frigeri, A., Nicchia, G. P., Svelto, M., Chiesa, G., Sirtori, C., Scanziani, E., Rizzo, C., et al. (2006). Effects of chronic treatment with statins and fenofibrate on rat skeletal muscle: A biochemical, histological and electrophysiological study. *Br. J. Pharmacol.* **149,** 909–919.

Plassart-Schiess, E., Gervais, A., Eymard, B., Lagueny, A., Pouget, J., Warter, J. M., Fardeau, M., Jentsch, T. J., and Fontaine, B. (1998). Novel muscle chloride channel (CLCN1) mutations in myotonia congenita with various modes of inheritance including incomplete dominance and penetrance. *Neurology* **50,** 1176–1179.

Pusch, M., Steinmeyer, K., and Jentsch, T. J. (1994). Low single channel conductance of the major skeletal muscle chloride channel, ClC-1. *Biophys. J.* **66,** 149–152.

Pusch, M., Steinmeyer, K., Koch, M. C., and Jentsch, T. J. (1995). Mutations in dominant human myotonia congenita drastically alter the voltage dependence of the ClC-1 chloride channel. *Neuron* **15,** 1455–1463.

Pusch, M., Liantonio, A., Bertorello, L., Accardi, A., De Luca, A., Pierno, S., Tortorella, V., and Camerino, D. C. (2000). Pharmacological characterization of chloride channels belonging to the ClC family by the use of chiral clofibric acid derivatives. *Mol. Pharmacol.* **58,** 498–507.

Rhodes, T. H., Vite, C. H., Giger, U., Patterson, D. F., Fahlke, C., and George, A. L., Jr. (1999). A missense mutation in canine ClC-1 causes recessive myotonia congenita in the dog. *FEBS Lett.* **456,** 54–58.

Ricker, K., Moxley, R. T., 3rd, Heine, R., and Lehmann-Horn, F. (1994). Myotonia fluctuans. A third type of muscle sodium channel disease. *Arch. Neurol.* **51,** 1095–1102.

Rogers, C. S., Vanoye, C. G., Sullenger, B. A., and George, A. L., Jr. (2002). Functional repair of a mutant chloride channel using a trans-splicing ribozyme. *J. Clin. Invest.* **110,** 1783–1789.

Rüdel, R., Dengler, R., Ricker, K., Haass, A., and Emser, W. (1980). Improved therapy of myotonia with the lidocaine derivative tocainide. *J. Neurol.* **222,** 275–278.

Rüdel, R., Ricker, K., and Lehmann-Horn, F. (1988). Transient weakness and altered membrane characteristic in recessive generalized myotonia (Becker). *Muscle Nerve* **11,** 202–211.

Rychkov, G. Y., Pusch, M., Roberts, M. L., Jentsch, T. J., and Bretag, A. H. (1998). Permeation and block of the skeletal muscle chloride channel, ClC-1, by foreign anions. *J. Gen. Physiol.* **111**, 653–665.

Sangiuolo, F., Botta, A., Mesoraca, A., Servidei, S., Merlini, L., Fratta, G., Novelli, G., and Dallapiccola, B. (1998). Identification of five new mutations and three novel polymorphisms in the muscle chloride channel gene (CLCN1) in 20 Italian patients with dominant and recessive myotonia congenita. Mutations in brief no. 118. Online. *Hum. Mutat.* **11**, 331.

Sasaki, R., Ichiyasu, H., Ito, N., Ikeda, T., Takano, H., Ikeuchi, T., Kuzuhara, S., Uchino, M., Tsuji, S., and Uyama, E. (1999). Novel chloride channel gene mutations in two unrelated Japanese families with Becker's autosomal recessive generalized myotonia. *Neuromuscul. Disord.* **9**, 587–592.

Sasaki, R., Ito, N., Shimamura, M., Murakami, T., Kuzuhara, S., Uchino, M., and Uyama, E. (2001). A novel CLCN1 mutation: P480T in a Japanese family with Thomsen's myotonia congenita. *Muscle Nerve* **24**, 357–363.

Saviane, C., Conti, F., and Pusch, M. (1999). The muscle chloride channel ClC-1 has a double-barreled appearance that is differentially affected in dominant and recessive myotonia. *J. Gen. Physiol.* **113**, 457–468.

Savitha, M. R., Krishnamurthy, B., Hyderi, A., Farhan Ul, H., and Ramachandra, N. B. (2006). (2006). Myotonia congenita—A successful response to carbamazepine. *Indian J. Pediatr.* **73**, 431–433.

Sejersen, T., Anvret, M., and George, A. L. (1996). Autosomal recessive myotonia congenita with missense mutation (T550M) in chloride channel gene (CLCN1). *Neuromuscul. Disord. Suppl.* **S47.**

Simpson, B. J., Height, T. A., Rychkov, G. Y., Nowak, K. J., Laing, N. G., Hughes, B. P., and Bretag, A. H. (2004). Characterization of three myotonia-associated mutations of the CLCN1 chloride channel gene via heterologous expression. *Hum. Mutat.* **24**, 185.

Sloan-Brown, K., and George, A. L., Jr. (1997). Inheritance of three distinct muscle chloride channel gene (CLCN1) mutations in a single recessive myotonia congenita family. *Neurology* **48**, 542–543.

Sonoda, Y., Gotow, T., Kuriyama, M., Nakahara, K., Arimura, K., and Osame, M. (1994). Electrical myotonia of rabbit skeletal muscles by HMG-CoA reductase inhibitors. *Muscle Nerve* **17**, 891–897.

Speedy, H. (1990). Exaggerated physiological responses to propofol in myotonic dystrophy. *Br. J. Anaesth.* **64**, 110–112.

Steinmeyer, K., Klocke, R., Ortland, C., Gronemeier, M., Jockusch, H., Grunder, S., and Jentsch, T. J. (1991a). Inactivation of muscle chloride channel by transposon insertion in myotonic mice. *Nature* **354**, 304–308.

Steinmeyer, K., Ortland, C., and Jentsch, T. J. (1991b). Primary structure and functional expression of a developmentally regulated skeletal muscle chloride channel. *Nature* **354**, 301–304.

Steinmeyer, K., Lorenz, C., Pusch, M., Koch, M. C., and Jentsch, T. J. (1994). Multimeric structure of ClC-1 chloride channel revealed by mutations in dominant myotonia congenita (Thomsen). *EMBO J.* **13**, 737–743.

Streib, E. W. (1987). AAEE minimonograph #27: Differential diagnosis of myotonic syndromes. *Muscle Nerve* **10**, 603–615.

Sun, C., Tranebjaerg, L., Torbergsen, T., Holmgren, G., and Van Ghelue, M. (2001). Spectrum of CLCN1 mutations in patients with myotonia congenita in Northern Scandinavia. *Eur. J. Hum. Genet.* **9**, 903–909.

Thomsen, J. (1876). Tonische Krämpfe in willkürlich beweglichen Muskeln in Folge von ererbter psychischer Disposition (Ataxia muscularis?). *Arch. Psychiatr. Nervenkr.* **6**, 702–718.

Thomson, G. W. (1956). Quinidine as a cause of sudden death. *Circulation* **14**, 757–765.

Trip, J., Drost, G., van Engelen, B. G., and Faber, C. G. (2006). Drug treatment for myotonia. *Cochrane Database Syst Rev.* CD004762.

Trip, J., Drost, G., Verbove, D. J., van der Kooi, A. J., Kuks, J. B., Notermans, N. C., Verschuuren, J. J., de Visser, M., van Engelen, B. G., Faber, C. G., and Ginjaar, I. B. (2008). In tandem analysis of CLCN1 and SCN4A greatly enhances mutation detection in families with non-dystrophic myotonia. *Eur. J. Hum. Genet.* **16,** 921–929.

Van Beekvelt, M. C., Drost, G., Rongen, G., Stegeman, D. F., Van Engelen, B. G., and Zwarts, M. J. (2006). Na^+-K^+-ATPase is not involved in the warming-up phenomenon in generalized myotonia. *Muscle Nerve* **33,** 514–523.

Wagner, S., Deymeer, F., Kurz, L. L., Benz, S., Schleithoff, L., Lehmann-Horn, F., Serdaroglu, P., Ozdemir, C., and Rüdel, R. (1998). The dominant chloride channel mutant G200R causing fluctuating myotonia: Clinical findings, electrophysiology, and channel pathology. *Muscle Nerve* **21,** 1122–1128.

Wakeman, B., Babu, D., Tarleton, J., and Macdonald, I. M. (2008). Extraocular muscle hypertrophy in myotonia congenita. *J. AAPOS* **12,** 294–296.

Watts, R. L., Watkins, J., and Watts, D. C. (1978). A new mouse mutant with abnormal muscle function: Comparison with the Re-dy mouse. *In* "The Biochemistry of Myasthenia Gravis and Muscular Dystrophy" (R. M. Marchbanks and G. G. Lunt, eds.), pp. 331–334. Academic Press, London.

Westphal, C. (1883). Demonstration zweier Fälle von Thomsen'scher Krankheit. *Berl. Klin. Wschr.* **20,** 153.

White, J. B., and Plaskett, J. (1904). "Nervous", "stiff-legged" or "fainting goats". *Am. Vet. Rev.* **28,** 556.

Wu, F. F., Ryan, A., Devaney, J., Warnstedt, M., Korade-Mirnics, Z., Poser, B., Escriva, M. J., Pegoraro, E., Yee, A. S., Felice, K. J., Giuliani, M. J., and Mayer, R. F. (2002). Novel CLCN1 mutations with unique clinical and electrophysiological consequences. *Brain* **125,** 2392–2407.

Zhang, J., George, A. L., Jr., Griggs, R. C., Fouad, G. T., Roberts, J., Kwiecinski, H., Connolly, A. M., and Ptacek, L. J. (1996). Mutations in the human skeletal muscle chloride channel gene (CLCN1) associated with dominant and recessive myotonia congenita. *Neurology* **47,** 993–998.

Zhang, J., Sanguinetti, M. C., Kwiecinski, H., and Ptacek, L. J. (2000). Mechanism of inverted activation of ClC-1 channels caused by a novel myotonia congenita mutation. *J. Biol. Chem.* **275,** 2999–3005.

3

Familial Hemiplegic Migraine

Curtis F. Barrett,*,† **Arn M. J. M. van den Maagdenberg,***,†
Rune R. Frants,† **and Michel D. Ferrari***

*Department of Neurology, Leiden University Medical Center, Leiden,
The Netherlands
†Department of Human Genetics, Leiden University Medical Center, Leiden,
The Netherlands

I. Introduction
   A. Migraine is a disabling episodic disorder
   B. Epidemiology of migraine
   C. Comorbidity
II. The Migraine Attack: Clinical Phases and Pathophysiology
III. The Migraine Aura and Cortical Spreading Depression
   A. Cortical spreading depression in experimental animals
   B. CSD and the migraine aura
   C. CSD as the trigger of the headache phase
IV. Migraine as a Genetic Disorder
   A. Genetic epidemiology of migraine
   B. Environmental influences in migraine susceptibility
   C. The road to migraine gene discovery
V. Familial Hemiplegic Migraine: A Model for Common Migraine
   A. The FHM1 *CACNA1A* gene
   B. The FHM2 *ATP1A2* gene
   C. The FHM3 *SCN1A* gene
   D. Sporadic hemiplegic migraine (SHM)
   E. FHM4 and beyond

Advances in Genetics, Vol. 63
Copyright 2008, Elsevier Inc. All rights reserved.

0065-2660/08 $35.00
DOI: 10.1016/S0065-2660(08)01003-1

VI. Functional Consequences of FHM Mutations
    A. $Ca_V2.1$ (P/Q-type) calcium channel (FHM1)
    B. Na^+/K^+-ATPase $\alpha2$ transporter (FHM2)
    C. $Na_V1.1$ (FHM3)
VII. FHM as an Ionopathy: Identifying a Common
    Theme among FHM Subtypes
    A. FHM1
    B. FHM2
    C. FHM3
VIII. Concluding Remarks
    Acknowledgments
    References

ABSTRACT

Migraine is a severely debilitating episodic disorder affecting up to 12% of the general population. Migraine arises from both genetic and environmental factors, complicating our understanding of what makes the migraine brain susceptible to attacks. In recent years, powerful genetic screening tools have revealed several single genes linked to migraine. One example of a monogenic subtype of migraine is familial hemiplegic migraine (FHM), a rare form of migraine with aura. The fact that FHM and common multifactorial migraine have many overlapping clinical features indicates that they likely share underlying pathophysiological pathways. In addition, the identification of monogenic subtypes has made it possible to generate suitable animal models for migraine. The purpose of this review is to present an overview of the clinical features of migraine and discuss the continuing highway of migraine gene discovery. The genes associated with FHM will be discussed, including what we have learned from studying the functional consequences of FHM mutations in cellular and animal models. © 2008, Elsevier Inc.

I. INTRODUCTION

A. Migraine is a disabling episodic disorder

Migraine is one of the most common neurological disorders, affecting at least 12% of the general population. A primary headache disorder, migraine is typically characterized by disabling episodic attacks of severe throbbing unilateral headache, often accompanied by any or all of the following: nausea, hypersensitivity to sound and/or light, and head movement. A typical episode can last about a day, and the frequency of attacks varies widely among migraineurs (see below).

In approximately one third of migraineurs, the headache is preceded by an aura that typically lasts 20–60 min. This aura usually consists of unilateral homonymous visual symptoms, such as flashing lights in a jagged pattern and vision loss (hemianopsia), which begin paracentrally and slowly expand over minutes as a hemifield defect. Migraine auras may also include other transient focal neurological symptoms. Because of the wide range of symptoms exhibited by migraineurs, the International Headache Society has designated diagnostic criteria for defining migraine, as shown in Table 3.1 (Headache Classification Committee of The International Headache Society, 2004).

Migraine can be classified into two general categories: (1) migraine *with aura* [MA; previously called classic(al) migraine], in which at least some of the attacks temporally coincide with distinct transient focal neurological aura symptoms; and (2) migraine *without aura* (MO; previously called common migraine), in which there are no identifiable associated neurological symptoms of a focal nature (Silberstein *et al.*, 2002). In both groups, migraine attacks are often precipitated by triggers specific to the individual, and can include hormonal changes, certain odors, flashing lights, sound, dehydration, alcohol consumption, and altered sleep and eating patterns. However, the trigger can also vary within individuals and can change during their lifetime.

B. Epidemiology of migraine

The onset of migraine attacks can occur at any age, but rarely beyond the age of 50 years. In females, the peak incidence is from 12 to 17 years of age. In males, the incidence of migraine peaks several years earlier than females, from 5 to 11 years of age (Haut *et al.*, 2006). Thus, in children migraine is more prevalent in boys, while in the general population, the overall prevalence is two-thirds female. Peak prevalence is around 40 years of age (Lipton *et al.*, 2001; Scher *et al.*, 1999). The median attack frequency is 18 migraine attacks per year but can vary widely, ranging from one or two in a lifetime to as frequent as several times per month; about 10% of migraineurs have attacks at least once weekly (Goadsby *et al.*, 2002). Migraine is rated by the World Health Organization as among the most disabling chronic disorders (Menken *et al.*, 2000). As such, migraine is estimated to be the most costly neurological disorder: in the European Community (Andlin-Sobocki *et al.*, 2005) and the United States (Stewart *et al.*, 2003), annual migraine costs total more than €27 billion and $19.6 billion, respectively.

Table 3.1. Diagnostic Criteria for Migraine and Familial Hemiplegic Migraine

A. Migraine diagnostic criteria
   At least five attacks fulfilling the following criteria:
   (a) Headache attacks lasting 4–72 h (untreated or unsuccessfully treated)
   (b) Headache attacks have at least two of the following characteristics
       • Unilateral location
       • Pulsating quality
       • Moderate or severe pain intensity
       • Aggravated by movement or disrupting routine physical activity
   (c) During the headache at least one of the following
       • Nausea and/or vomiting
       • Photophobia and phonophobia
   (d) Not attributed to any other disorder
B. Familial hemiplegic migraine diagnostic criteria
   [a]At least two attacks fulfilling the following criteria:
   (a) Aura consisting of fully reversible motor weakness and at least one of the
       following
       • Fully reversible visual symptoms including positive (e.g., flickering
         lights) and/or negative (i.e., loss of vision) features
       • Fully reversible sensory symptoms including positive (i.e., "pins and
         needles" sensation) and/or negative (i.e., numbness) features
       • Fully reversible dysphasic speech disturbances
   (b) At least two of the following
       • At least one aura symptom that develops over ≥5 min and/or different
         aura symptoms occurring in succession over ≥5 min
       • Each aura symptom lasts 5 min or more but less than 24 h
       • Headache fulfilling the criteria for migraine without aura begins
         during the aura or follows the aura within 60 min
   (c) At least one first- or second-degree relative meeting the criteria above
   (d) Not attributed to any other disorder

From the International Headache Society.
[a]Sporadic hemiplegic migraine follows the same criteria, with the exception of family history (c).

C. Comorbidity

Migraine is often associated with other episodic brain disorders (Goadsby *et al.*, 2002); in addition, comorbidity is often bidirectional. Together, this suggests common underlying pathology and genetic origin. Migraineurs have significantly increased risk (several-fold higher than nonmigraineurs) for epilepsy (Haut *et al.*, 2006; Ludvigsson *et al.*, 2006), depression and anxiety disorders

(Breslau and Davis, 1993; Breslau et al., 2003; Radat and Swendsen, 2005), patent foramen ovale (Bousser and Welch, 2005), and stroke. The incidence of comorbidity can depend on age and cofactors including smoking and the use of oral contraceptives (Bousser and Welch, 2005). Moreover, high-frequency migraineurs have a 16-fold increase in white matter and cerebellar lesions (Kruit et al., 2004).

II. THE MIGRAINE ATTACK: CLINICAL PHASES AND PATHOPHYSIOLOGY

Migraine attacks consist of up to four distinct phases (Blau, 1992), described briefly below. It is important to note that not every patient will experience every phase:

1. *Premonitory phase.* Up to one third of patients may occasionally experience premonitory symptoms prior to the onset of the aura and/or headache phase. These warning signs may include changes in mood (e.g., depression or irritation), hyperactivity or fatigue, neck pain, smell anomalies, food cravings, and water retention (Giffin et al., 2003).
2. *Aura phase.* As many as one third of patients may experience transient visual, sensory, motor, brainstem, or cognitive aura symptoms in at least some of their attacks. These focal neurological symptoms typically last up to an hour, but may last from several hours to several days (Russell and Olesen, 1996). In addition, the aura phase typically precedes the headache phase by 5 min to 1 h, but can also occur simultaneously with, or following, the headache phase.
3. *Headache phase.* The headache phase itself is highly debilitating, with severe throbbing pain and associated symptoms including nausea, vomiting, and sensitivity to light and sound as well as to head movement and touch. In most cases, this phase lasts for about a day, but can last 3 days.
4. *Recovery and interictal phase.* Recovery from a migraine attack can take from several hours to several days (Giffin et al., 2005; Kelman, 2006), during which the patient is often fatigued. Following recovery, the patient enters an interictal period in which clinical symptoms are absent.

III. THE MIGRAINE AURA AND CORTICAL SPREADING DEPRESSION

A. Cortical spreading depression in experimental animals

It is now generally accepted that the aura preceding a migraine headache is not the result of vascular constriction, as was previously believed, but rather is most likely caused by the human equivalent of a phenomenon called cortical spreading depression (CSD) (Eikermann-Haerter and Moskowitz, 2008; Haerter *et al.*, 2005; Lauritzen, 1994). First reported by Leão in the 1940s (Leão, 1944a, b), CSD can be best described as a wave of transient ($\sim$1–2 min) depolarization that sweeps in all directions across the cerebral cortex. This depolarization is believed to involve both neurons and glial cells, and spreads through contiguous areas of brain cortex at a rate of 2–5 mm/min without regard to functional cortical divisions or arterial territories. In experimental animals, the electrophysiological changes are associated with changes in cerebral blood flow (CBF): there is a small, brief reduction in CBF followed by a large increase in CBF lasting several minutes, after which blood flow is reduced for up to an hour and is accompanied by a loss of the cerebrovascular response to hypercapnia (Piper *et al.*, 1991). CSD is currently believed to serve to protect the brain from strong stimuli, as it can be readily triggered by direct electrical stimulation of the cerebral cortex, cerebrocortical trauma or ischemia, or by cortical application of strongly depolarizing concentrations of K^+ or excitatory amino acids such as glutamate (Somjen, 2001).

B. CSD and the migraine aura

A wealth of clinical evidence supports the notion that CSD is the underlying basis of the migraine aura. For example, when mapped to the visual cortex, visual aura symptoms typically travel from the center of the visual field to the periphery at a speed of approximately 3 mm/min (Lashley, 1941), on par with the propagation rate of CSD in experimental animals. In addition, the positive (e.g., scintillations, paresthesias) and negative (e.g., scotomata, paresis) phenomena of the migraine aura can be well explained by the biphasic nature of CSD, in which the neuronal depression is frequently preceded by transient hyperexcitability. Most importantly however, functional neuroimaging studies have demonstrated that the changes in CBF during a migraine aura are highly reminiscent of those observed in experimental animals during CSD. Using functional MRI, Hadjikhani and colleagues (2001) found a focal increase in the blood oxygen level-dependent (BOLD) signal that spread through the occipital cortex at a rate of 3.5 mm/min. The direction and speed of the spread were in agreement with the visual experiences reported by the patient, and the increased BOLD signal was followed by a decrease in signal. This pattern would be consistent with a brief initial rise in CBF followed by a longer-lasting decrease in blood flow.

C. CSD as the trigger of the headache phase

Despite mounting evidence that CSD underlies the migraine aura, it remains unclear whether CSD may trigger the headache phase itself, perhaps via activation of the trigeminovascular system. Animal experiments (for review, see

Eikermann-Haerter and Moskowitz, 2008; Haerter et al., 2005) have provided evidence in support of this hypothesis. For example, high K^+-induced CSD in the rat parietal cortex can activate ipsilateral trigeminal nucleus caudalis neurons and cause long-lasting elevated blood flow in the middle meningeal artery, as well as dural plasma protein leakage that can be inhibited by ipsilateral trigeminal nerve resection (Bolay et al., 2002). Moreover, treatment with several classes of migraine prophylactic drugs inhibits experimentally induced CSDs as measured with electrophysiology (Ayata et al., 2006) and cerebral blood flow (Akerman and Goadsby, 2005).

In contrast to animal studies, the connection between CSD and the headache phase in patients remains an open question. Goadsby (2001) reviewed the mainly clinical arguments against the hypothesis that CSD may also trigger headache mechanisms. First, only one third of migraineurs report having auras, raising the question of how the headache phase is triggered in patients without aura. One possible explanation is that all migraineurs might indeed have SD, but that perhaps MO patients exhibit (C)SD in clinically silent subcortical areas of the brain without propagating to the visual cortex (Haerter et al., 2005). Although formally possible, this hypothesis is difficult to test directly; however, one case reported spreading cerebral hypoperfusion during the headache phase in an MO patient (Woods et al., 1994). Second, the aura and the headache phase would be predicted to occur on opposite sides if CSD activates the trigemino-vascular system, but in some cases the headache can occur on the same side as the aura. Third, in some cases aura can occur *after* the headache has started, suggesting that it might not serve to trigger the headache. Fourth, aura is not unique to migraineurs. For example, auras have been reported with attacks of cluster headache (Bahra et al., 2002; Silberstein et al., 2000), paroxysmal hemicrania (Matharu and Goadsby, 2001), and hemicrania continua (Peres et al., 2002). Finally, treatment with intranasal ketamine has been reported to abort migraine aura but failed to prevent the headache phase (Kaube et al., 2000). Yet despite these arguments, the exact role, if any, that spreading depression plays in triggering migraine headache mechanisms remains unclear and the subject of intense investigation.

IV. MIGRAINE AS A GENETIC DISORDER

A. Genetic epidemiology of migraine

There is considerable evidence to support the notion that migraine has an underlying genetic basis. First, migraine frequently runs in families (Kors et al., 2004), and population-based studies have confirmed that the risk of migraine in first-degree relatives is 1.5- to 4-fold greater than in nonrelated individuals. This familial risk is highest for MA patients, with a young age of onset and a high attack severity and disease disability (Russell and Olesen, 1995; Stewart *et al.*, 1997, 2006).

   Using the basis of differing estimates of heritability, some reports have suggested that MA and MO are different entities (Ludvigsson *et al.*, 2006; Russell and Olesen, 1995; Russell *et al.*, 2002). However, several arguments suggest instead that the aura component may have some heritable biological distinction. For example, many migraineurs experience both MA and MO attacks throughout their lifetime, for example, MA in childhood, MO in young adulthood, and MA again in later life. Furthermore, an Australian study of more than 6000 twin pairs used the basis of the patterns and severity of migraine symptoms to identify disease subtypes ("latent classes"), arguing against MA and MO being distinct disorders (Nyholt *et al.*, 2004). Indeed, a study of over 200 Finnish migraine families suggested that migraine is a continuum, with pure migraine (MO) at one end, MA at the other, and a combination of both in between (Kallela *et al.*, 2001). Thus, at present, the different migraine subtypes appear to be different clinical manifestations of the same disorder.

B. Environmental influences in migraine susceptibility

Studies of twin pairs are often the best method to gauge the relative contributions of genetic and environmental factors. In twin pairs drawn from the general population, the pairwise concordance rates for migraine were significantly higher among monozygotic than among dizygotic twin pairs, indicating that genetic factors are important in the susceptibility to migraine. However, as concordance is not 100%, environmental factors clearly play a role as well. Thus, migraine is truly a multifactorial complex disorder (Gervil *et al.*, 1999; Honkasalo *et al.*, 1995; Mulder *et al.*, 2003; Ulrich *et al.*, 1999). The relative importance of genetic factors can be gleaned from a large population-based study of 30,000 twin pairs from six countries, in which migraine heritability was 40–50%; in contrast, shared environmental factors were considered to play only a minor role in migraine susceptibility (Mulder *et al.*, 2003). This finding is supported by studies comparing twins raised either together or apart (Svensson *et al.*, 2003; Ziegler *et al.*, 1998).

C. The road to migraine gene discovery

The identification of genes involved in multifactorial disorders is often hampered by several complicating factors. First, because multiple genes contribute to disease susceptibility, each gene involved contributes a relatively small effect to the phenotype. Second, the phenotypic expression can be influenced by nongenetic, and often variable, intrinsic and extrinsic factors. Finally, complex disorders are often highly prevalent and may present at older ages, thus preventing a reliable distinction between affected and nonaffected populations.

To identify genes associated with migraine, several genetic approaches have been applied, the most successful of which has been the identification of single genes in families with monogenic subtypes of migraine. This approach is predicated on the assumption that rare monogenic subtypes and common multifactorial forms of migraine have common genes and related biochemical pathways for the trigger threshold and initiation mechanisms of migraine episodes. Thus, rare monogenic variants may serve as a suitable genetic and/or functional model for the common complex forms. In this regard, studying the functional consequences of the causative gene mutations might hint at pathogenic pathways common to all subtypes of migraine.

A second linkage analysis approach often used to study complex traits is affected sib-pair analysis, which identifies chromosomal regions shared by affected siblings at a higher probability than by chance alone. This is then followed by case-control association studies testing single nucleotide polymorphisms (SNPs) in candidate genes within the shared regions. The goal is to identify SNPs, and thus gene variants, that statistically differ in frequency between case subjects and controls and may underlie increased susceptibility to the disease.

A third, hypothesis-driven approach is to directly test candidate genes in case-control association studies. An interesting new twist to this approach will be the application of nonhypothesis-driven testing for genome-wide association by scanning hundreds of thousands of SNPs in extended and clinically homogeneous populations (Hirschhorn and Daly, 2005).

V. FAMILIAL HEMIPLEGIC MIGRAINE: A MODEL FOR COMMON MIGRAINE

Familial hemiplegic migraine (FHM) is a rare subtype of migraine with aura. The diagnostic criteria for FHM, as determined by the International Headache Society, are presented in Table 3.1. In brief, FHM is characterized by at least some degree of hemiparesis during the aura phase (Ferrari, 1998). This hemiparesis may last from minutes to several hours or even days and often patients are initially misdiagnosed with epilepsy. Apart from the hemiparesis, the other headache and aura features of the FHM attack are strikingly similar to those exhibited during common MA attacks. In addition to attacks with hemiparesis, the majority of FHM patients also experience attacks of "normal" migraine with or without aura (Ducros et al., 2001; Terwindt et al., 1998). Thus, from a clinical perspective, FHM seems a valid model for the common forms of migraine both with and without aura

(Ferrari, 1998). However, it bears mentioning that in 20% of cases, FHM may also be associated with cerebellar ataxia and other neurological symptoms such as epilepsy, mental retardation, brain edema, and (sometimes fatal) coma.

To date, three genes for FHM have been identified, and, based on unpublished linkage results in several families, there are likely more to come. These genes are discussed below.

A. The FHM1 *CACNA1A* gene

In 1996, the first gene linked to FHM was identified as the *CACNA1A* gene on chromosome 19p13 (Ophoff *et al.*, 1996). Responsible for approximately half of all families with FHM, the FHM1 gene encodes the ion-conducting, pore-forming α_{1A} subunit of $Ca_V 2.1$, a neuronal voltage-gated calcium channel. The main function of $Ca_V 2.1$ (P/Q-type) calcium channels is to trigger the release of neurotransmitters, both at central synapses and at the neuromuscular junction (Catterall, 1998). Over 50 mutations in *CACNA1A* have been associated with a wide range of clinical phenotypes (Adams and Snutch, 2007; Haan *et al.*, 2005; Piedras-Rentería *et al.*, 2007; Stam *et al.*, 2008; van den Maagdenberg *et al.*, 2007), including both pure FHM (Ophoff *et al.*, 1996) and combinations of FHM with various degrees of cerebellar ataxia (Ducros *et al.*, 2001; Marti *et al.*, 2008; Ophoff *et al.*, 1996) and fatal coma due to excessive cerebral edema (Kors *et al.*, 2001). Mutations in the *CACNA1A* gene have also been linked to disorders not associated with FHM, including episodic ataxia type 2 (EA2) (Jen *et al.*, 2004; Ophoff *et al.*, 1996), progressive ataxia (Yue *et al.*, 1997), spinocerebellar ataxia type 6 (Zhuchenko *et al.*, 1997), and both absence (Imbrici *et al.*, 2004) and generalized (Haan *et al.*, 2005; Jouvenceau *et al.*, 2001) epilepsy. Interestingly, in several FHM families, FHM1 *CACNA1A* mutations were also found in family members who did not exhibit FHM but instead exhibited "normal" nonparetic migraine, suggesting that gene mutations for FHM may also be responsible for the common forms of migraine, possibly due to the contributions of different genetic and nongenetic modulatory factors.

B. The FHM2 *ATP1A2* gene

The second gene identified for FHM does not code for an ion channel, but rather for a transporter that catalyzes the exchange of Na^+ and K^+ across the cell membrane. The *ATP1A2* gene on chromosome 1q23 encodes the $\alpha 2$ subunit of a Na^+/K^+-ATPase (De Fusco *et al.*, 2003; Marconi *et al.*, 2003). This enzyme utilizes ATP hydrolysis to transport Na^+ ions out of the cell in exchange for K^+ ions into the cell, thereby providing the steep Na^+ gradient essential for the transport of glutamate and Ca^{2+}. During development, *ATP1A2* is predominantly expressed in neurons, but expression shifts to glial cells by adulthood (McGrail *et al.*, 1991; Moseley *et al.*, 2003). In adults, an important function of this specific ATPase is to modulate the removal of potassium and glutamate from the synaptic cleft into the glial cell.

Mutations in the *ATP1A2* gene cause at least 20% of FHM cases and have been associated with pure FHM (De Fusco *et al.*, 2003; Fernandez *et al.*, 2008; Riant *et al.*, 2005; Vanmolkot *et al.*, 2006) as well as with FHM together with cerebellar ataxia (Spadaro *et al.*, 2004), alternating hemiplegia of childhood (Bassi *et al.*, 2004; Swoboda *et al.*, 2004), benign familial infantile convulsions (Vanmolkot *et al.*, 2003), and other forms of epilepsy (Haan *et al.*, 2005). Interestingly, in an Italian family, a variant in the *ATP1A2* gene segregated with basilar migraine, a subtype of MA characterized by aura symptoms attributable to the brainstem and both occipital lobes (Ambrosini *et al.*, 2005). However, no functional studies were reported, precluding a definite conclusion as to whether this gene variation is also *causally* linked to basilar migraine. It is also interesting to note that *ATP1A2* variants were identified in two non-FHM migraine families, suggesting that this gene may play an important role in the susceptibility to common forms of migraine (Todt *et al.*, 2005).

C. The FHM3 *SCN1A* gene

The most recent FHM gene identified is *SCN1A* on chromosome 2q24 (Dichgans *et al.*, 2005; Vanmolkot *et al.*, 2007; see also Gargus and Tournay, 2007). This gene encodes the α subunit of a neuronal $Na_V1.1$ sodium channel. Voltage-gated sodium channels are primarily responsible for the generation and propagation of action potentials in excitable cells (reviewed in Yu and Catterall, 2003). In the brain, $Na_V1.1$ channels are expressed primarily on inhibitory neurons (Ogiwara *et al.*, 2007), and their absence leads to severe ataxia and seizures, presumably due to increased neuronal excitability (Yu *et al.*, 2006). Different mutations in *SCN1A* are known to be associated with childhood epilepsy and febrile seizures (for review, see Meisler and Kearney, 2005).

D. Sporadic hemiplegic migraine (SHM)

Hemiplegic migraine patients are not always clustered in families. Sporadic cases in which the patient has no affected family members are often seen. These cases may indeed represent the first "FHM" patient (*de novo* mutation) in a family (Stam *et al.*, 2008; Thomsen *et al.*, 2003a). Like FHM, SHM has an overlapping clinical phenotype with the common forms of migraine; indeed, the diagnostic criteria for SHM are identical to FHM except for the presence of affected first- or second-degree relatives (see Table 3.1). In addition, SHM and normal migraine also show a remarkable genetic epidemiological relationship. SHM patients have a markedly increased risk of typical migraine with aura, and first-degree relatives have an increased risk of both MA and MO (Thomsen *et al.*, 2003b). Although an initial study reported CACNA1A mutations in only 2 of 27 SHM patients (Terwindt *et al.*, 2002), when this study was expanded, de Vries *et al.* (2007) found causal mutations in the FHM genes in a higher proportion of SHM patients, confirming a genetic relationship between FHM and SHM.

E. FHM4 and beyond

Given the relatively rapid path to the discovery of three genes linked to FHM, and continually improving sequencing and screening methods, new FHM genes will undoubtedly be discovered. This is supported by reports of FHM families without identified mutations in any of the three FHM genes (Vanmolkot *et al.*, 2007). Given the common pathway in which all three FHM proteins play a role (see Section VII), it is likely that new candidate genes will encode proteins that play key roles in neuronal excitability and/or ion homeostasis.

VI. FUNCTIONAL CONSEQUENCES OF FHM MUTATIONS

A. Ca$_V$2.1 (P/Q-type) calcium channel (FHM1)

Upon the initial discovery of four missense mutations in CACNA1A linked to FHM (Ophoff *et al.*, 1996), the path to understanding how the mutations give rise to the disease seemed relatively straightforward. However, this proved to be initially elusive, as functional analyses of these four mutations yielded conflicting results. In some cases, mutations gave rise to an apparent loss of function (Barrett *et al.*, 2005; Cao and Tsien, 2005; Cao *et al.*, 2004; Hans *et al.*, 1999; Kraus *et al.*, 1998; Tottene *et al.*, 2002), while in other cases, a gain-of-function phenotype was observed (Hans *et al.*, 1999; Kraus *et al.*, 1998; Melliti *et al.*, 2003; Tottene *et al.*, 2002). This was in stark contrast to CACNA1A mutations linked to EA2, in which all mutations studied confer a clear loss-of-function phenotype to the channel (Guida *et al.*, 2001; Imbrici *et al.*, 2004; Jen *et al.*, 2001; Jeng *et al.*, 2006; Jouvenceau *et al.*, 2001; Spacey *et al.*, 2004; Wan *et al.*, 2005; Wappl *et al.*, 2002).

To date, 19 FHM1 mutations have been identified in CACNA1A (Fig. 3.1), and most have been studied in heterologous expression systems (Barrett *et al.*, 2005; Jen *et al.*, 2001; Kraus *et al.*, 1998, 2000; Melliti *et al.*, 2003; Mullner *et al.*, 2004; Tottene *et al.*, 2002, 2005). As with the original four mutations, no clear pattern of either loss or gain of function initially emerged. Because of these conflicting results, researchers turned to expressing recombinant channels in primary cultured neurons from *Cacna1a* knockout mice lacking endogenous P/Q-type channels. Four such studies have been published (Cao and

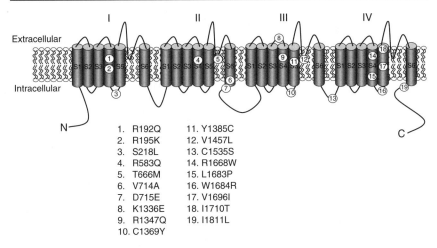

1.	R192Q	11. Y1385C
2.	R195K	12. V1457L
3.	S218L	13. C1535S
4.	R583Q	14. R1668W
5.	T666M	15. L1683P
6.	V714A	16. W1684R
7.	D715E	17. V1696I
8.	K1336E	18. I1710T
9.	R1347Q	19. I1811L
10.	C1369Y	

Figure 3.1. Schematic topographic drawing of the Ca$_V$2.1 (α_{1A}) subunit of the P/Q-type voltage-gated calcium channel encoded by the FHM1 gene *CACNA1A*. The protein consists of four repeating domains (I–IV), each containing six transmembrane helices (S1–S6). The approximate locations of mutations linked to FHM are indicated. Amino acid positions refer to GenBank accession number X99897.

Tsien, 2005; Cao *et al.*, 2004; Tottene *et al.*, 2002, 2005). In three studies (Cao and Tsien, 2005; Cao *et al.*, 2004; Tottene *et al.*, 2002), the mutations caused dramatically reduced whole-cell currents when compared with wild-type channels. In the fourth (Tottene *et al.*, 2005), the mutation gave rise to an increased current density at negative voltages. The functional consequences of the original four mutations on synaptic transmission in cultured neurons were also examined, and, consistent with decreased whole-cell current density, every mutation caused impaired synaptic transmission (Cao and Tsien, 2005; Cao *et al.*, 2004).

Despite the evidence for FHM1 mutations causing a loss-of-function phenotype, increasing evidence was emerging in support of gain-of-function effects. This evidence primarily arose from a series of elegant studies measuring the single-channel properties of channels bearing FHM mutations. A careful examination of channel open probability and unitary current amplitude over a range of voltages revealed that the mutant channels exhibit a negative shift in voltage dependence, meaning that the channels open more readily with less depolarization. This effect was also reflected in whole-cell currents measured in S218L channels expressed in *Cacna1a* knockout neurons, in which significantly larger currents were observed at negative voltages (Tottene *et al.*, 2005). In all, ten FHM1 mutations, including all four original mutations, are reported to activate at more negative voltages than wild-type channels (Hans *et al.*, 1999; Melliti *et al.*, 2003; Tottene *et al.*, 2002, 2005).

One important lesson learned from these studies is that reliable functional data cannot always be obtained solely using exogenously expressed channels. Many factors, including the origin species of the channel, the splice variant chosen, the accessory subunit(s), and the expression level can all influence the results and their interpretation. In addition, single-channel measurements (which were so effective in revealing gain-of-function effects for FHM mutations, as discussed above) are technically very challenging to perform and analyze. To help overcome these factors, the mutation can be introduced into the orthologous gene in the mouse genome by homologous recombination (the so-called gene-targeting strategy, and more specifically the knockin strategy). The advantages of this approach are that the mutation is present in all relevant splice forms of the gene, and the mutant protein is expressed in the appropriate tissues, at the appropriate levels, at the appropriate developmental stages, and with its appropriate binding partners. This strategy was employed for the R192Q mutation, which neutralizes a positive charge in the first voltage sensor (van den Maagdenberg et al., 2004) (see Fig. 3.1). In principle, one would predict this mutation to decrease voltage sensitivity. However, a detailed examination of the channel properties revealed a clear gain of function: despite being expressed at levels equivalent to wild-type channels, much larger currents were observed at negative voltages (at −30 mV, the current through mutant channels was better than three times that of wild-type channels); at positive voltages, mutant

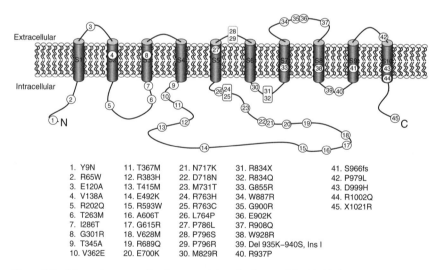

1. Y9N	11. T367M	21. N717K	31. R834X	41. S966fs
2. R65W	12. R383H	22. D718N	32. R834Q	42. P979L
3. E120A	13. T415M	23. M731T	33. G855R	43. D999H
4. V138A	14. E492K	24. R763H	34. W887R	44. R1002Q
5. R202Q	15. R593W	25. R763C	35. G900R	45. X1021R
6. T263M	16. A606T	26. L764P	36. E902K	
7. I286T	17. G615R	27. P786L	37. R908Q	
8. G301R	18. V628M	28. P796S	38. W928R	
9. T345A	19. R689Q	29. P796Q	39. Del 935K–940S, Ins I	
10. V362E	20. E700K	30. M829R	40. R937P	

Figure 3.2. Schematic topographic drawing of the α2 subunit of the Na⁺/K⁺-ATPase transporter encoded by the FHM2 gene *ATP1A2*. The protein contains 10 transmembrane helices (S1–S10). The approximate locations of mutations linked to FHM are indicated. Amino acid positions refer to GenBank accession number NM_000702.

channels carried the same current as wild-type channels. Thus, the increased currents at negative voltages could be explained in full by a negative shift in activation for the mutant channels. This shift is consistent with similar effects observed for other FHM1 mutations.

Consistent with the role of $Ca_V2.1$ channels in mediating transmitter release at the neuromuscular junction, the R192Q mutant mice exhibited increased transmission from motor neurons. This increase was reflected in both a higher frequency of miniature (spontaneous) endplate potentials and larger evoked endplate potentials, both of which showed a gene-dosage effect (Kaja et al., 2005; van den Maagdenberg et al., 2004). Interestingly, the effect of the mutation on evoked endplate potentials was observed only when extracellular Ca^{2+} was reduced to subphysiological levels, suggesting a ceiling effect in physiological Ca^{2+} concentrations, and perhaps explaining why R192Q patients do not present with chronic neuromuscular deficits.

As discussed in Section III, CSD is believed to be the physiological correlate of the migraine aura (Haerter et al., 2005; Lauritzen, 1994). The naturally occurring mouse *Cacna1a* mutants *tottering* and *leaner* exhibit decreased susceptibility for CSD, manifesting as an increased threshold for inducing a CSD event (Ayata et al., 2000). Functionally, these mutations convey clear electro-physiological loss-of-function effects to the channels (Dove et al., 1998; Wakamori et al., 1998). In contrast, in keeping with a gain-of-function effect in FHM1, the R192Q knockin mice exhibited a decreased threshold for inducing a CSD event (van den Maagdenberg et al., 2004). Together, these findings suggest a direct positive correlation between channel function and spreading depression: more Ca^{2+} influx leads to an increased susceptibility for spreading depression in the cortex. This may have clinical significance, as decreasing calcium channel activity to wild-type levels in the brain might help rescue the migraine brain.

B. Na⁺/K⁺-ATPase α2 transporter (FHM2)

To date, 45 mutations in the *ATP1A2* gene have been linked to FHM (Fig. 3.2), and several have been studied functionally. *In vitro* assays revealed that mutant proteins have decreased transporter activity, ranging from no activity at all (Capendeguy and Horisberger, 2004; Koenderink et al., 2005) to decreased affinity for K^+, and/or decreased catalytic turnover (De Fusco et al., 2003; Segall et al., 2004, 2005). Thus, the mutation leads to reduced reuptake of K^+ and glutamate from the synaptic cleft into glial cells.

Because the FHM2 mutations impair protein function, it is interesting to compare the FHM2 phenotype with that of mice lacking *Atp1a2* expression. Two lines of knockout mice have been generated (Ikeda et al., 2004; James et al., 1999), and in both cases the mice die immediately at birth due to severe motor deficits and failing respiration. Examination of the developing embryos revealed neuronal loss in the amygdala and piriform cortex secondary to neural hyperactivity (Ikeda et al., 2003). Moreover, Atp1a2 knockout mice on a 129sv genetic background survive briefly (<1 day) but display frequent generalized seizures (Ikeda et al., 2004). This finding is particularly interesting, as clinically ATP1A2 mutations can cause epilepsy. Heterozygous Atp1a2$^{+/-}$ mice are viable but exhibit enhanced fear and anxiety following conditioned fear stimuli; this is likely due to neuronal hyperactivity in the amygdala and piriform cortex (Ikeda et al., 2003).

Currently, no Atp1a2 knockin model of FHM2 is available. However, certain predictions can be made with regard to possible effects on cortical excitability. Given the loss-of-function phenotype identified by functional in vitro assays, one would expect increased neuronal excitability. In addition, because FHM2 is clinically indistinguishable from FHM1, it would be interesting to measure susceptibility to CSD in FHM2 mice.

C. Na$_V$1.1 (FHM3)

Unlike CACNA1A and ATP1A2, in which a plethora of FHM mutations have been identified, to date only two mutations in SCN1A have been linked to FHM (Dichgans et al., 2005; Vanmolkot et al., 2007) (Fig. 3.3). The first mutation identified, Q1489K, has been functionally studied (Dichgans et al., 2005). However, because of well-known SCN1A cDNA instability in bacteria during cloning procedures, the mutation was instead introduced into the homologous Na$_V$1.5 channel, where it acted to accelerate recovery from inactivation. Given

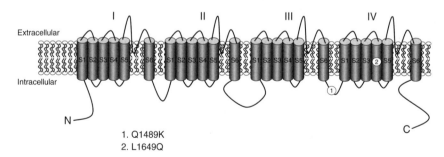

1. Q1489K
2. L1649Q

Figure 3.3. Schematic topographic drawing of the Na$_V$1.1 subunit of the voltage-gated sodium channel encoded by the FHM3 gene SCN1A. The protein contains four repeating domains (I–IV), each consisting of six transmembrane helices (S1–S6). The approximate locations of mutations linked to FHM are indicated. Amino acid positions refer to GenBank accession number AB093548.

that voltage-gated sodium channels play a critical role in initiation and propagation of action potentials, accelerated recovery from inactivation would likely cause increased neuronal firing.

The complete picture is not likely to be so simple, however. Studies have shown that, unlike Na$_V$1.5, Na$_V$1.1 is expressed primarily on *inhibitory* neurons (Ogiwara *et al.*, 2007; Yu *et al.*, 2006), where increased firing would lead to decreased neuronal excitability. In addition, we can infer the possible outcome of FHM3 mutations in Na$_V$1.1 by considering other missense mutations. For example, *de novo* loss-of-function mutations in *SCN1A* have been linked to severe myoclonic epilepsy in infancy (SMEI) (Claes *et al.*, 2001; Fujiwara *et al.*, 2003; Ohmori *et al.*, 2002). It was initially unclear why decreased function of a channel that drives neuronal excitability would lead to increased excitability, but this apparent paradox was resolved when the *Scn1a* gene was disrupted by gene targeting to generate a mouse model for SMEI (Yu *et al.*, 2006). Heterozygous *Scn1a*$^{+/-}$ mice exhibited reduced sodium currents in their inhibitory neurons, leading to hyperexcitability. A subsequent knockin mouse model for SMEI showing epileptic activity confirmed these results (Ogiwara *et al.*, 2007).

The true test of the effect of FHM3 mutations will await either *in vitro* functional assays in Na$_V$1.1 or the generation of an FHM3 knockin mouse model. Until then, the available knockout mouse may serve as a suitable model to examine FHM and common migraine.

VII. FHM AS AN IONOPATHY: IDENTIFYING A COMMON THEME AMONG FHM SUBTYPES

The present challenge is in understanding how mutations in three different genes encoding three different proteins can cause similar (or even identical) clinical outcomes. Currently, FHM (and perhaps migraine in general) can best be described as a generalized hyperexcitability due to impaired ion homeostasis. Figure 3.4 shows a simplified schematic of a synapse between an excitable presynaptic neuron and a postsynaptic neuron, as well as input from an inhibitory interneuron. Both the localization of the FHM protein and the nature of the defect will determine the net effect of the mutation on overall brain activity.

A. FHM1

Ca$_V$2.1 is depicted at the presynaptic terminal of excitable neurons, where a gain of function will increase transmitter release. It bears mentioning, however, that this same channel is also present at nerve terminals of inhibitory neurons, where the FHM1 mutation may serve to dampen excitability, perhaps partially protecting the brain during an episodic attack.

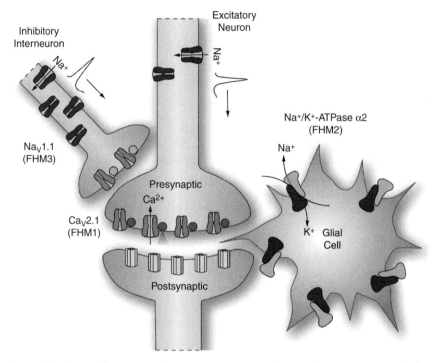

Figure 3.4. Cartoon depicting a glutamatergic synapse in the central nervous system and the functional roles of proteins encoded by the FHM1, FHM2, and FHM3 genes. $Ca_V2.1$ calcium channels are located in the presynaptic terminal of excitatory and inhibitory neurons. In response to an invading action potential, these channels gate, allowing Ca^{2+} to enter and triggering vesicle fusion and glutamate release into the synaptic cleft. K^+ in the synaptic cleft is removed in part by the action of the Na^+/K^+-ATPase located at the surface of glial cells (astrocytes). Removing extracellular K^+ serves to dampen neuronal excitability and generates a Na^+ gradient, which drives uptake of glutamate from the cleft by transporters, for example, EAAT1. Lastly, the $Na_V1.1$ voltage-gated sodium channel is expressed in inhibitory interneurons, where they serve to initiate and propagate action potentials. Gain-of-function mutations in $Ca_V2.1$ and loss-of-function mutations in the ATPase and $Na_V1.1$ will each lead to a net effect of increased general excitability.

B. FHM2

Early in development, the Na^+/K^+-ATPase $\alpha2$ subunit is expressed in neurons (McGrail *et al.*, 1991; Moseley *et al.*, 2003). However, in adult brain, expression is predominantly in glial cells, which play a critical role in removing transmitters and K^+ from the synaptic cleft. Therefore, loss-of-function mutations in this protein will likely lead to increased excitability. Support for this comes from a mutation in *SLC1A3* in a sporadic patient with migraine and alternating hemiplegia (Jen et al., 2005). SLC1A3 codes for the EAAT1 glutamate transporter expressed in glial cells (Banner et al., 2002; Kawakami et al., 1994), and functional studies of the mutant protein revealed dramatically impaired glutamate transport (Jen et al., 2005). Moreover, knocking down endogenous EAAT1 levels with antisense oligonucleotides caused increased extracellular glutamate levels and excitotoxicity (Rothstein et al., 1996).

C. FHM3

$Na_V1.1$ is expressed primarily in inhibitory interneurons (Ogiwara et al., 2007; Yu et al., 2006), where a loss of function (e.g., as seen with SMEI) would dampen inhibitory input, again leading to overall increased excitability.

VIII. CONCLUDING REMARKS

Migraine is a multifactorial episodic disorder of the brain affecting a large percentage of the general population. The high prevalence, combined with the highly disabling nature of the disease, causes migraine to rank as high impact on the lives of patients and their families, as well as society as a whole. Although many fundamental questions remain unanswered, the work of many investigators has greatly expanded our understanding of the triggers and underlying mechanisms of the migraine attack.

The identification of FHM genes has allowed the development of genetic screening methods for diagnosing familial and sporadic hemiplegic migraine. In addition, employing the power of gene-targeting methods has yielded mouse models of FHM, and these mice have proven valuable in the study of the physiology of migraine, and will undoubtedly help guide research in the development of migraine prevention and treatment.

Acknowledgments

We are grateful to Boukje de Vries and Kaate Vanmolkot for helpful discussions and ion channel schematics, and to Daniela Pietrobon for critically commenting on sections of the manuscript. M. D. F. is the recipient of a Netherlands Organization for Scientific Research (NWO) Vici grant (nr. 918.56.602). Additional support was provided by the Centre for Medical Systems Biology (CMSB) in the framework of the Netherlands Genomics Initiative (NGI), and from the Eurohead Project.

References

Adams, P. J., and Snutch, T. P. (2007). Calcium channelopathies: Voltage-gated calcium channels. Subcell Biochem. **45**, 215–251.
Akerman, S., and Goadsby, P. J. (2005). Topiramate inhibits cortical spreading depression in rat and cat: Impact in migraine aura. NeuroReport **16**, 1383–1387.

Ambrosini, A., D'Onofrio, M., Grieco, G. S., Di Mambro, A., Fortini, D., Nicoletti, F., Nappi, G., Sances, G., Schoenen, J., Buzzi, M. G., Santorelli, F. M., and Pierelli, F. (2005). A new mutation on the ATPA2 gene in one Italian family with basilar-type migraine linked to the FHM2 locus. *Neurology* **64,** A132.

Andlin-Sobocki, P., Jonsson, B., Wittchen, H. U., and Olesen, J. (2005). Cost of disorders of the brain in Europe. *Eur. J. Neurol.* **12,** 1–27.

Ayata, C., Shimizu-Sasamata, M., Lo, E. H., Noebels, J. L., and Moskowitz, M. A. (2000). Impaired neurotransmitter release and elevated threshold for cortical spreading depression in mice with mutations in the alpha 1A subunit of P/Q type calcium channels. *Neuroscience* **95,** 639–645.

Ayata, C., Jin, H., Kudo, C., Dalkara, T., and Moskowitz, M. A. (2006). Suppression of cortical spreading depression in migraine prophylaxis. *Ann. Neurol.* **59,** 652–661.

Bahra, A., May, A., and Goadsby, P. J. (2002). Cluster headache: A prospective clinical study in 230 patients with diagnostic implications. *Neurology* **58,** 354–361.

Banner, S. J., Fray, A. E., Ince, P. G., Steward, M., Cookson, M. R., and Shaw, P. J. (2002). The expression of the glutamate re-uptake transporter excitatory amino acid transporter 1 (EAAT1) in the normal human CNS and in motor neurone disease: An immunohistochemical study. *Neuroscience* **109,** 27–44.

Barrett, C. F., Cao, Y. Q., and Tsien, R. W. (2005). Gating deficiency in a familial hemiplegic migraine type 1 mutant P/Q-type calcium channel. *J. Biol. Chem.* **280,** 24064–24071.

Bassi, M. T., Bresolin, N., Tonelli, A., Nazos, K., Crippa, F., Baschirotto, C., Zucca, C., Bersano, A., Dolcetta, D., Boneschi, F. M., Barone, V., and Casari, G. (2004). A novel mutation in the ATP1A2 gene causes alternating hemiplegia of childhood. *J. Med. Genet.* **41,** 621–628.

Blau, J. N. (1992). Migraine: Theories of pathogenesis. *Lancet* **339,** 1202–1207.

Bolay, H., Reuter, U., Dunn, A. K., Huang, Z., Boas, D. A., and Moskowitz, M. A. (2002). Intrinsic brain activity triggers trigeminal meningeal afferents in a migraine model. *Nat. Med.* **8,** 136–142.

Bousser, M. G., and Welch, K. M. (2005). Relation between migraine and stroke. *Lancet Neurol.* **4,** 533–542.

Breslau, N., and Davis, G. C. (1993). Migraine, physical health and psychiatric disorder: A prospective epidemiological study in young adults. *J. Psychiatr. Res.* **27,** 211–221.

Breslau, N., Lipton, R. B., Stewart, W. F., Schultz, L. R., and Welch, K. M. (2003). Comorbidity of migraine and depression: Investigating potential etiology and prognosis. *Neurology* **60,** 1308–1312.

Cao, Y. Q., and Tsien, R. W. (2005). Effects of familial hemiplegic migraine type 1 mutations on neuronal P/Q-type Ca2+ channel activity and inhibitory synaptic transmission. *Proc. Natl Acad. Sci. USA* **102,** 2590–2595.

Cao, Y. Q., Piedras-Renteria, E. S., Smith, G. B., Chen, G., Harata, N. C., and Tsien, R. W. (2004). Presynaptic Ca^{2+} channels compete for channel type-preferring slots in altered neurotransmission arising from Ca^{2+} channelopathy. *Neuron* **43,** 387–400.

Capendeguy, O. O., and Horisberger, J. D. (2004). Functional effects of Na+, K+ ATPase gene mutations linked to familial hemiplegic migraine. *Neuromol. Med.* **6,** 105–116.

Catterall, W. A. (1998). Structure and function of neuronal Ca2+ channels and their role in neurotransmitter release. *Cell Calcium* **24,** 307–323.

Claes, L., Del-Favero, J., Ceulemans, B., Lagae, L., Van Broeckhoven, C., and De Jonghe, P. (2001). De novo mutations in the sodium-channel gene SCN1A cause severe myoclonic epilepsy of infancy. *Am. J. Hum. Genet.* **68,** 1327–1332.

De Fusco, M., Marconi, R., Silvestri, L., Atorino, L., Rampoldi, L., Morgante, L., Ballabio, A., Aridon, P., and Casari, G. (2003). Haploinsufficiency of ATP1A2 encoding the Na^{+}/K^{+} pump $\alpha 2$ subunit associated with familial hemiplegic migraine type 2. *Nat. Genet.* **33,** 192–196.

de Vries, B., Freilinger, T., Vanmolkot, K. R., Koenderink, J. B., Stam, A. H., Terwindt, G. M., Babini, E., van den Boogerd, E. H., van den Heuvel, J. J., Frants, R. R., Haan, J., Pusch, M., van den Maagdenberg, A. M., Ferrari, M. D., and Dichgans, M. (2007). Systematic analysis of three FHM genes in 39 sporadic patients with hemiplegic migraine. *Neurology* **69,** 2170–2176.

Dichgans, M., Freilinger, T., Eckstein, G., Babini, E., Lorenz-Depiereux, B., Biskup, S., Ferrari, M. D., Herzog, J., van den Maagdenberg, A. M., Pusch, M., and Strom, T. M. (2005). Mutation in the neuronal voltage-gated sodium channel SCN1A causes familial hemiplegic migraine. Lancet 366, 371–377.

Dove, L. S., Abbott, L. C., and Griffit, W. H. (1998). Whole-cell and single-channel analysis of P-type calcium currents in cerebellar purkinje cells of leaner mutant mice. J. Neurosci. 18, 7687–7699.

Ducros, A., Denier, C., Joutel, A., Cecillon, M., Lescoat, C., Vahedi, K., Darcel, F., Vicaut, E., Bousser, M. G., and Tournier-Lasserve, E. (2001). The clinical spectrum of familial hemiplegic migraine associated with mutations in a neuronal calcium channel. N. Engl. J. Med. 345, 17–24.

Eikermann-Haerter, K., and Moskowitz, M. A. (2008). Animal models of migraine headache and aura. Curr. Opin. Neurol. 21, 294–300.

Fernandez, D. M., Hand, C. K., Sweeney, B. J., and Parfrey, N. A. (2008). A novel ATP1A2 gene mutation in an Irish familial hemiplegic migraine kindred. Headache 48, 101–108.

Ferrari, M. D. (1998). Migraine. Lancet 351, 1043–1051.

Fujiwara, T., Sugawara, T., Mazaki-Miyazaki, E., Takahashi, Y., Fukushima, K., Watanabe, M., Hara, K., Morikawa, T., Yagi, K., Yamakawa, K., and Inoue, Y. (2003). Mutations of sodium channel alpha subunit type 1 (SCN1A) in intractable childhood epilepsies with frequent generalized tonic-clonic seizures. Brain 126, 531–546.

Gargus, J. J., and Tournay, A. (2007). Novel mutation confirms seizure locus SCN1A is also familial hemiplegic migraine locus FHM3. Pediatr. Neurol. 37, 407–410.

Gervil, M., Ulrich, V., Kaprio, J., Olesen, J., and Russell, M. B. (1999). The relative role of genetic and environmental factors in migraine without aura. Neurology 53, 995–999.

Giffin, N. J., Ruggiero, L., Lipton, R. B., Silberstein, S., Tvedskov, J. F., Olesen, J., Altman, J., Goadsby, P. J., and Macrae, A. (2003). Premonitory symptoms in migraine: An electronic diary study. Neurology 60, 935–940.

Giffin, N. J., Lipton, R. B., Silberstein, S. D., Tvedskov, J. F., Olesen, J., and Goadsby, P. J. (2005). The migraine postdrome: An electronic diary study. Cephalalgia 25, 958.

Goadsby, P. J. (2001). Migraine, aura and cortical spreading depression: Why are we still talking about it? Ann. Neurol. 49, 4–6.

Goadsby, P. J., Lipton, R. B., and Ferrari, M. D. (2002). Migraine—Current understanding and treatment. N. Engl. J. Med. 346, 257–270.

Guida, S., Trettel, F., Pagnutti, S., Mantuano, E., Tottene, A., Veneziano, L., Fellin, T., Spadaro, M., Stauderman, K., Williams, M., Volsen, S., Ophoff, R., Frants, R., Jodice, C., Frontali, M., and Pietrobon, D. (2001). Complete loss of P/Q calcium channel activity caused by a CACNA1A missense mutation carried by patients with episodic ataxia type 2. Am. J. Hum. Genet. 68, 759–764.

Haan, J., Kors, E. E., Vanmolkot, K. R., van den Maagdenberg, A. M., Frants, R. R., and Ferrari, M. D. (2005). Migraine genetics: An update. Curr. Pain Headache Rep. 9, 213–220.

Hadjikhani, N., Sanchez del Rio, M., Wu, O., Schwartz, D., Bakker, D., Fischl, B., Wong, K. K., Cutrer, F. M., Rosen, B. R., Tootell, R. B. H., Sorensen, A. G., and Moskowitz, M. A. (2001). Mechanisms of migraine aura revealed by functional MRI in human visual cortex. Proc. Natl Acad. Sci. USA 98, 4687–4692.

Haerter, K., Ayata, C., and Moskowitz, M. A. (2005). Cortical spreading depression: A model for understanding migraine biology and future drug targets. Headache Currents 2, 97–103.

Hans, M., Luvisetto, S., Williams, M. E., Spagnolo, M., Urrutia, A., Tottene, A., Brust, P. F., Johnson, E. C., Harpold, M. M., Stauderman, K. A., and Pietrobon, D. (1999). Functional consequences of mutations in the human alpha(1A) calcium channel subunit linked to familial hemiplegic migraine. J. Neurosci. 19, 1610–1619.

Haut, S. R., Bigal, M. E., and Lipton, R. B. (2006). Chronic disorders with episodic manifestations: Focus on epilepsy and migraine. *Lancet Neurol.* **5**, 148–157.

Headache Classification Committee of The International Headache Society (2004). The International Classification of Headache Disorders (second edition). *Cephalalgia* **24**, 1–160.

Hirschhorn, J. N., and Daly, M. J. (2005). Genome-wide association studies for common diseases and complex traits. *Nat. Rev. Genet.* **6**, 95–108.

Honkasalo, M. L., Kaprio, J., Winter, T., Heikkilä, K., Sillanpää, M., and Koskenvuo, M. (1995). Migraine and concomitant symptoms among 8167 adult twin pairs. *Headache* **35**, 70–78.

Ikeda, K., Onaka, T., Yamakado, M., Nakai, J., Ishikawa, T. O., Taketo, M. M., and Kawakami, K. (2003). Degeneration of the amygdala/piriform cortex and enhanced fear/anxiety behaviors in sodium pump alpha2 subunit (Atp1a2)-deficient mice. *J. Neurosci.* **23**, 4667–4676.

Ikeda, K., Onimaru, H., Yamada, J., Inoue, K., Ueno, S., Onaka, T., Toyoda, H., Arata, A., Ishikawa, T. O., Taketo, M. M., Fukuda, A., and Kawakami, K. (2004). Malfunction of respiratory-related neuronal activity in Na+, K+-ATPase alpha2 subunit-deficient mice is attributable to abnormal Cl$^-$ homeostasis in brainstem neurons. *J. Neurosci.* **24**, 10693–10701.

Imbrici, P., Jaffe, S. L., Eunson, L. H., Davies, N. P., Herd, C., Robertson, R., Kullmann, D. M., and Hanna, M. G. (2004). Dysfunction of the brain calcium channel Ca$_V$2.1 in absence epilepsy and episodic ataxia. *Brain* **127**, 2682–2692.

James, P. F., Grupp, I. L., Grupp, G., Woo, A. L., Askew, G. R., Croyle, M. L., Walsh, R. A., and Lingrel, J. B. (1999). Identification of a specific role for the Na,K-ATPase alpha 2 isoform as a regulator of calcium in the heart. *Mol. Cell* **3**, 555–563.

Jen, J., Wan, J., Graves, M., Yu, H., Mock, A. F., Coulin, C. J., Kim, G., Yue, Q., Papazian, D. M., and Baloh, R. W. (2001). Loss-of-function EA2 mutations are associated with impaired neuromuscular transmission. *Neurology* **57**, 1843–1848.

Jen, J., Kim, G. W., and Baloh, R. W. (2004). Clinical spectrum of episodic ataxia type 2. *Neurology* **62**, 17–22.

Jen, J. C., Wan, J., Palos, T. P., Howard, B. D., and Baloh, R. W. (2005). Mutation in the glutamate transporter EAAT1 causes episodic ataxia, hemiplegia, and seizures. *Neurology* **65**, 529–534.

Jeng, C. J., Chen, Y. T., Chen, Y. W., and Tang, C. Y. (2006). Dominant-negative effects of human P/Q-type Ca^{2+} channel mutations associated with episodic ataxia type 2. *Am. J. Physiol. Cell Physiol.* **290**, 1209–1220.

Jouvenceau, A., Eunson, L. H., Spauschus, A., Ramesh, V., Zuberi, S. M., Kullmann, D. M., and Hanna, M. G. (2001). Human epilepsy associated with dysfunction of the brain P/Q-type calcium channel. *Lancet* **358**, 801–807.

Kaja, S., van de Ven, R. C., Broos, L. A., Veldman, H., van Dijk, J. G., Verschuuren, J. J., Frants, R. R., Ferrari, M. D., van den Maagdenberg, A. M., and Plomp, J. J. (2005). Gene dosage-dependent transmitter release changes at neuromuscular synapses of CACNA1A R192Q knockin mice are non-progressive and do not lead to morphological changes or muscle weakness. *Neuroscience* **135**, 81–95.

Kallela, M., Wessman, M., Havanka, H., Palotie, A., and Farkkila, M. (2001). Familial migraine with and without aura: Clinical characteristics and co-occurrence. *Eur. J. Neurol.* **8**, 441–449.

Kaube, H., Herzog, J., Kaufer, T., Dichgans, M., and Diener, H. C. (2000). Aura in some patients with familial hemiplegic migraine can be stopped by intranasal ketamine. *Neurology* **55**, 139–141.

Kawakami, H., Tanaka, K., Nakayama, T., Inoue, K., and Nakamura, S. (1994). Cloning and expression of a human glutamate transporter. *Biochem. Biophys. Res. Commun.* **199**, 171–176.

Kelman, L. (2006). The postdrome of the acute migraine attack. *Cephalalgia* **26**, 214–220.

Koenderink, J. B., Zifarelli, G., Qiu, L. Y., Schwarz, W., De Pont, J. J. H. H. M., Bamberg, E., and Friedrich, T. (2005). Na,K-ATPase mutations in familial hemiplegic migraine lead to functional inactivation. *Biochim. Biophys. Acta* **1669**, 61–68.

Kors, E. E., Terwindt, G. M., Vermeulen, F. L. M. G., Fitzsimons, R. B., Jardine, P. E., Heywood, P., Love, S., van den Maagdenberg, A. M. J. M., Haan, J., Frants, R. R., and Ferrari, M. D. (2001). Delayed cerebral edema and fatal coma after minor head trauma: Role of CACNA1A calcium channel subunit gene and relationship with familial hemiplegic migraine. *Ann. Neurol.* **49,** 753–760.

Kors, E. E., Vanmolkot, K. R., Haan, J., Frants, R. R., van den Maagdenberg, A. M., and Ferrari, M. D. (2004). Recent findings in headache genetics. *Curr. Opin. Neurol.* **17,** 283–288.

Kraus, R. L., Sinnegger, M. J., Glossmann, H., Hering, S., and Striessnig, J. (1998). Familial hemiplegic migraine mutations change alpha(1A) Ca^{2+} channel kinetics. *J. Biol. Chem.* **273,** 5586–5590.

Kraus, R. L., Sinnegger, M. J., Koschak, A., Glossmann, H., Stenirri, S., Carrera, P., and Striessnig, J. (2000). Three new familial hemiplegic migraine mutants affect P/Q-type Ca(2+) channel kinetics. *J. Biol. Chem.* **275,** 9239–9243.

Kruit, M. C., van Buchem, M. A., Hofman, P. A., Bakkers, J. T., Terwindt, G. M., Ferrari, M. D., and Launer, L. J. (2004). Migraine as a risk factor for subclinical brain lesions. *J. Am. Med. Assoc.* **291,** 427–434.

Lashley, K. S. (1941). Patterns of cerebral integration indicated by the scotomas of migraine. *Arch. Neurol. Psychiatry* **46,** 331–339.

Lauritzen, M. (1994). Pathophysiology of the migraine aura. The spreading depression theory. *Brain* **117,** 199–210.

Leão, A. A. P. (1944a). Further observations on the spreading depression of activity in the cerebral cortex. *J. Neurophysiol.* **10,** 409–414.

Leão, A. A. P. (1944b). Spreading depression of activity in the cerebral cortex. *J. Neurophysiol.* **7,** 359–390.

Lipton, R. B., Stewart, W. F., Diamond, S., Diamond, M. L., and Reed, M. (2001). Prevalence and burden of migraine in the United States: Data from the American Migraine Study II. *Headache* **41,** 646–657.

Ludvigsson, P., Hesdorffer, D., Olafsson, E., Kjartansson, O., and Hauser, W. A. (2006). Migraine with aura is a risk factor for unprovoked seizures in children. *Ann. Neurol.* **59,** 210–213.

Marconi, R., De Fusco, M., Aridon, P., Plewnia, K., Rossi, M., Carapelli, S., Ballabio, A., Morgante, L., Musolino, R., Epifanio, A., Micieli, G., De Michele, G., and Casari, G. (2003). Familial hemiplegic migraine type 2 is linked to 0.9Mb region on chromosome 1q23. *Ann. Neurol.* **53,** 376–381.

Marti, S., Baloh, R. W., Jen, J. C., Straumann, D., and Jung, H. H. (2008). Progressive cerebellar ataxia with variable episodic symptoms—Phenotypic diversity of R1668W CACNA1A mutation. *Eur. Neurol.* **60,** 16–20.

Matharu, M. S., and Goadsby, P. J. (2001). Post-traumatic chronic paroxysmal hemicrania (CPH) with aura. *Neurology* **56,** 273–275.

McGrail, K. M., Phillips, J. M., and Sweadner, K. J. (1991). Immunofluorescent localization of three Na,K-ATPase isozymes in the rat central nervous system: Both neurons and glia can express more than one Na,K-ATPase. *J. Neurosci.* **11,** 381–391.

Meisler, M. H., and Kearney, J. A. (2005). Sodium channel mutations in epilepsy and other neurological disorders. *J. Clin. Invest.* **115,** 2010–2017.

Melliti, K., Grabner, M., and Seabrook, G. R. (2003). The familial hemiplegic migraine mutation R192Q reduces G-protein-mediated inhibition of P/Q-type (Ca(V)2.1) calcium channels expressed in human embryonic kidney cells. *J. Physiol.* **546,** 337–347.

Menken, M., Munsat, T. L., and Toole, J. F. (2000). The global burden of disease study—Implications for neurology. *Arch. Neurol.* **57,** 418–420.

Moseley, A. E., Lieske, S. P., Wetzel, R. K., James, P. F., He, S., Shelly, D. A., Paul, R. J., Boivin, G. P., Witte, D. P., Ramirez, J. M., Sweadner, K. J., and Lingrel, J. B. (2003). The Na, K-ATPase alpha 2 isoform is expressed in neurons, and its absence disrupts neuronal activity in newborn mice. *J. Biol. Chem.* **278,** 5317–5324.

Mulder, E. J., Van Baal, C., Gaist, D., Kallela, M., Kaprio, J., Svensson, D. A., Nyholt, D. R., Martin, N. G., MacGregor, A. J., Cherkas, L. F., Boomsma, D. I., and Palotie, A. (2003). Genetic and environmental influences on migraine: A twin study across six countries. *Twin Res.* **6,** 422–431.

Mullner, C., Broos, L. A., van den Maagdenberg, A. M., and Striessnig, J. (2004). Familial hemiplegic migraine type 1 mutations K1336E, W1684R, and V1696I alter Cav2.1 Ca^{2+} channel gating: Evidence for beta-subunit isoform-specific effects. *J. Biol. Chem.* **279,** 51844–51850.

Nyholt, D. R., Gillespie, N. G., Heath, A. C., Merikangas, K. R., Duffy, D. L., and Martin, N. G. (2004). Latent class and genetic analysis does not support migraine with aura and migraine without aura as separate entities. *Genet. Epidemiol.* **26,** 231–244.

Ogiwara, I., Miyamoto, H., Morita, N., Atapour, N., Mazaki, E., Inoue, I., Takeuchi, T., Itohara, S., Yanagawa, Y., Obata, K., Furuichi, T., Hensch, T. K., and Yamakawa, K. (2007). Na(v)1.1 localizes to axons of parvalbumin-positive inhibitory interneurons: A circuit basis for epileptic seizures in mice carrying an Scn1a gene mutation. *J. Neurosci.* **27,** 5903–5914.

Ohmori, I., Ouchida, M., Ohtsuka, Y., Oka, E., and Shimizu, K. (2002). Significant correlation of the SCN1A mutations and severe myoclonic epilepsy in infancy. *Biochem. Biophys. Res. Commun.* **295,** 17–23.

Ophoff, R. A., Terwindt, G. M., Vergouwe, M. N., van Eijk, R., Oefner, P. J., Hoffman, S. M. G., Lamerdin, J. E., Mohrenweiser, H. W., Bulman, D. E., Ferrari, M., Haan, J., Lindhout, D., van Ommen, G. J. B., Hofker, M. H., Ferrari, M. D., and Frants, R. R. (1996). Familial hemiplegic migraine and episodic ataxia type-2 are caused by mutations in the Ca^{2+} channel gene CACNL1A4. *Cell* **87,** 543–552.

Peres, M. F. P., Siow, H. C., and Rozen, T. D. (2002). Hemicrania continua with aura. *Cephalalgia* **22,** 246–248.

Piedras-Rentería, E. S., Barrett, C. F., Cao, Y. Q., and Tsien, R. W. (2007). Voltage-gated calcium channels, calcium signaling, and channelopathies. *In* "Calcium: A Matter of Life or Death" (J. Krebs and M. Michalek, eds.), pp. 127–166. Elsevier, New York.

Piper, R. D., Lambert, G. A., and Duckworth, J. W. (1991). Cortical blood flow changes during spreading depression in cats. *Am. J. Physiol.* **261,** H96–H102.

Radat, F., and Swendsen, J. (2005). Psychiatric comorbidity in migraine: A review. *Cephalalgia* **25,** 165–178.

Riant, F., De Fusco, M., Aridon, P., Ducros, A., Ploton, C., Marchelli, F., Maciazek, J., Bousser, M. G., Casari, G., and Tournier-Lasserve, E. (2005). ATP1A2 mutations in 11 families with familial hemiplegic migraine. *Hum. Mutat.* **26,** 281.

Rothstein, J. D., Dykes-Hoberg, M., Pardo, C. A., Bristol, L. A., Jin, L., Kuncl, R. W., Kanai, Y., Hediger, M. A., Wang, Y., Schielke, J. P., and Welty, D. F. (1996). Knockout of glutamate transporters reveals a major role for astroglial transport in excitotoxicity and clearance of glutamate. *Neuron* **16,** 675–686.

Russell, M. B., and Olesen, J. (1995). Increased familial risk and evidence of genetic factor in migraine. *Br. Med. J.* **311,** 541–544.

Russell, M. B., and Olesen, J. (1996). A nosographic analysis of the migraine aura in a general population. *Brain* **119,** 355–361.

Russell, M. B., Ulrich, V., Gervil, M., and Olesen, J. (2002). Migraine without aura and migraine with aura are distinct disorders. A population-based twin survey. *Headache* **42,** 332–336.

Scher, A. I., Stewart, W. F., and Lipton, R. B. (1999). Migraine and headache: A meta-analytic approach. *In* "Epidemiology of Pain" (I. K. Crombie, ed.), pp. 159–170. IASP Press, Seattle, WA.

Segall, L., Scanzano, R., Kaunisto, M. A., Wessman, M., Palotie, A., Gargus, J. J., and Blostein, R. (2004). Kinetic alterations due to a missense mutation in the Na,K-ATPase alpha2 subunit cause familial hemiplegic migraine type 2. *J. Biol. Chem.* **279,** 43692–43696.

Segall, L., Mezzetti, A., Scanzano, R., Gargus, J. J., Purisima, E., and Blostein, R. (2005). Alterations in the alpha2 isoform of Na,K-ATPase associated with familial hemiplegic migraine type 2. *Proc. Natl Acad. Sci. USA* **102,** 11106–11111.

Silberstein, S. D., Niknam, R., Rozen, T. D., and Young, W. B. (2000). Cluster headache with aura. *Neurology* **54,** 219–221.

Silberstein, S. D., Lipton, R. B., and Goadsby, P. J. (2002). "Headache in Clinical Practice." Martin Dunitz, London.

Somjen, G. G. (2001). Mechanisms of spreading depression and hypoxic spreading depression-like depolarization. *Physiol. Rev.* **81,** 1065–1096.

Spacey, S. D., Hildebrand, M. E., Materek, L. A., Bird, T. D., and Snutch, T. P. (2004). Functional implications of a novel EA2 mutation in the P/Q-type calcium channel. *Ann. Neurol.* **56,** 213–220.

Spadaro, M., Ursu, S., Lehmann-Horn, F., Veneziano, L., Antonini, G., Giunti, P., Frontali, M., and Jurkat-Rott, K. (2004). A G301R Na^{+}/K^{+}-ATPase mutation causes familial hemiplegic migraine type 2 with cerebellar signs. *Neurogenetics* **5,** 177–185.

Stam, A. H., van den Maagdenberg, A. M., Haan, J., Terwindt, G. M., and Ferrari, M. D. (2008). Genetics of migraine: An update with special attention to genetic comorbidity. *Curr. Opin. Neurol.* **21,** 288–293.

Stewart, W. F., Staffa, J., Lipton, R. B., and Ottman, R. (1997). Familial risk of migraine: A population-based study. *Ann. Neurol.* **41,** 166–172.

Stewart, W. F., Ricci, J. A., Chee, E., Morganstein, D., and Lipton, R. (2003). Lost productive time and cost due to common pain conditions in the US workforce. *JAMA* **290,** 2443–2454.

Stewart, W. F., Bigal, M. E., Kolodner, K., Dowson, A., Liberman, J. N., and Lipton, R. B. (2006). Familial risk of migraine: Variation by proband age at onset and headache severity. *Neurology* **66,** 344–348.

Svensson, D. A., Larsson, B., Waldenlind, E., and Pedersen, N. L. (2003). Shared rearing environment in migraine: Results from twins reared apart and twins reared together. *Headache* **43,** 235–244.

Swoboda, K. J., Kanavakis, E., Xaidara, A., Johnson, J. E., Leppert, M. F., Schlesinger-Massart, M. B., Ptacek, L. J., Silver, K., and Youroukos, S. (2004). Alternating hemiplegia of childhood or familial hemiplegic migraine? A novel ATP1A2 mutation. *Ann. Neurol.* **55,** 884–887.

Terwindt, G. M., Ophoff, R. A., Haan, J., Vergouwe, M. N., van Eijk, R., Frants, R. R., and Ferrari, M. D. (1998). Variable clinical expression of mutations in the P/Q-type calcium channel gene in familial hemiplegic migraine. *Neurology* **50,** 1105–1110.

Terwindt, G., Kors, E., Haan, J., Vermeulen, F., Van den Maagdenberg, A., Frants, R., and Ferrari, M. (2002). Mutation analysis of the CACNA1A calcium channel subunit gene in 27 patients with sporadic hemiplegic migraine. *Arch. Neurol.* **59,** 1016–1018.

Thomsen, L. L., Ostergaard, E., Olesen, J., and Russell, M. B. (2003a). Evidence for a separate type of migraine with aura: Sporadic hemiplegic migraine. *Neurology* **60,** 595–601.

Thomsen, L. L., Ostergaard, E., Romer, S. F., Andersen, I., Eriksen, M. K., Olesen, J., and Russell, M. B. (2003b). Sporadic hemiplegic migraine is an aetiologically heterogeneous disorder. *Cephalalgia* **23,** 921–928.

Todt, U., Dichgans, M., Jurkat-Rott, K., Heinze, A., Zifarelli, G., Koenderink, J. B., Goebel, I., Zumbroich, V., Stiller, A., Ramirez, A., Friedrich, T., Gobel, H., and Kubisch, C. (2005). Rare missense variants in ATP1A2 in families with clustering of common forms of migraine. *Hum. Mutat.* **26,** 315–321.

Tottene, A., Tottene, A., Fellin, T., Pagnutti, S., Luvisetto, S., Striessnig, J., Fletcher, C., and Pietrobon, D. (2002). Familial hemiplegic migraine mutations increase Ca^{2+} influx through single human CaV2.1 channels and decrease maximal $Ca_V2.1$ current density in neurons. *Proc. Natl Acad. Sci. USA* **99**, 13284–13289.

Tottene, A., Pivotto, F., Fellin, T., Cesetti, T., van den Maagdenberg, A. M., and Pietrobon, D. (2005). Specific kinetic alterations of human CaV2.1 calcium channels produced by mutation S218L causing familial hemiplegic migraine and delayed cerebral edema and coma after minor head trauma. *J. Biol. Chem.* **280**, 17678–17686.

Ulrich, V., Gervil, M., Kyvik, K. O., Olesen, J., and Russell, M. B. (1999). Evidence of a genetic factor in migraine with aura: A population based Danish twin study. *Ann. Neurol.* **45**, 242–246.

van den Maagdenberg, A. M. J. M., Pietrobon, D., Pizzorusso, T., Kaja, S., Broos, L. A. M., Cesetti, T., van de Ven, R. C. G., Tottene, A., van der Kaa, J., Plomp, J. J., Frants, R. R., and Ferrari, M. D. (2004). A Cacna1a knock-in migraine mouse model with increased susceptibility to cortical spreading depression. *Neuron* **41**, 701–710.

van den Maagdenberg, A. M., Haan, J., Terwindt, G. M., and Ferrari, M. D. (2007). Migraine: Gene mutations and functional consequences. *Curr. Opin. Neurol.* **20**, 299–305.

Vanmolkot, K. R., Kors, E. E., Hottenga, J. J., Terwindt, G. M., Haan, J., Hoefnagels, W. A., Black, D. F., Sandkuijl, L. A., Frants, R. R., Ferrari, M. D., and van den Maagdenberg, A. M. (2003). Novel mutations in the Na^+, K^+-ATPase pump gene ATP1A2 associated with familial hemiplegic migraine and benign familial infantile convulsions. *Ann. Neurol.* **54**, 360–366.

Vanmolkot, K. R., Kors, E. E., Turk, U., Turkdogan, D., Keyser, A., Broos, L. A., Kia, S. K., van den Heuvel, J. J., Black, D. F., Haan, J., Frants, R. R., Barone, V., Ferrari, M. D., Casari, G., Koenderink, J. B., and van den Maagdenberg, A. M. (2006). Two *de novo* mutations in the Na, K-ATPase gene ATP1A2 associated with pure familial hemiplegic migraine. *Eur. J. Hum. Genet.* **14**, 555–560.

Vanmolkot, K. R., Babini, E., de Vries, B., Stam, A. H., Freilinger, T., Terwindt, G. M., Norris, L., Haan, J., Frants, R. R., Ramadan, N. M., Ferrari, M. D., Pusch, M., van den Maagdenberg, A. M., and Dichgans, M. (2007). The novel p.L1649Q mutation in the SCN1A epilepsy gene is associated with familial hemiplegic migraine: Genetic and functional studies. Mutation in brief #957. Online. *Hum. Mutat.* **28**, 522.

Wakamori, M., Yamazaki, K., Matsunodaira, H., Teramoto, T., Tanaka, I., Niidome, T., Sawada, K., Nishizawa, Y., Sekiguchi, N., Mori, E., Mori, Y., and Imoto, K. (1998). Single tottering mutations responsible for the neuropathic phenotype of the P-Type calcium channel. *J. Biol. Chem.* **273**, 34857–34867.

Wan, J., Khanna, R., Sandusky, M., Papazian, D. M., Jen, J. C., and Baloh, R. W. (2005). CACNA1A mutations causing episodic and progressive ataxia alter channel trafficking and kinetics. *Neurology* **64**, 2090–2097.

Wappl, E., Koschak, A., Poteser, M., Sinnegger, M. J., Walter, D., Eberhart, A., Groschner, K., Glossmann, H., Kraus, R. L., Grabner, M., and Striessnig, J. (2002). Functional consequences of P/Q-type Ca^{2+} channel Cav2.1 missense mutations associated with episodic ataxia type 2 and progressive ataxia. *J. Biol. Chem.* **277**, 6960–6966.

Woods, R. P., Iacoboni, M., and Mazziotta, J. C. (1994). Bilateral spreading cerebral hypoperfusion during spontaneous migraine headache. *N. Engl. J. Med.* **331**, 1689–1692.

Yu, F. H., and Catterall, W. A. (2003). Overview of the voltage-gated sodium channel family. *Genome Biol.* **4**, 207.

Yu, F. H., Mantegazza, M., Westenbroek, R. E., Robbins, C. A., Kalume, F., Burton, K. A., Spain, W. J., McKnight, G. S., Scheuer, T., and Catterall, W. A. (2006). Reduced sodium current in GABAergic interneurons in a mouse model of severe myoclonic epilepsy in infancy. *Nat. Neurosci.* **9**, 1142–1149.

Yue, Q., Jen, J. C., Nelson, S. F., and Baloh, R. W. (1997). Progressive ataxia due to a missense mutation in a calcium-channel gene. *Am. J. Hum. Genet.* **61,** 1078–1087.

Zhuchenko, O., Bailey, J., Donnen, P., Ashizawa, T., Stockton, D. W., Amos, C., Dobyns, W. B., Subramony, S. H., Zoghbi, H. Y., and Lee, C. C. (1997). Autosomal dominant cerebellar ataxia (SCA6) associated with small polyglutamine expansions in the α1A-voltage-dependent calcium channel. *Nat. Genet.* **15,** 62–69.

Ziegler, D. K., Hur, Y. M., Bouchard, T. J., Jr, Hassanein, R. S., and Barter, R. (1998). Migraine in twins raised together and apart. *Headache* **38,** 417–422.

4

Genetics and Molecular Pathophysiology of $Na_v1.7$-Related Pain Syndromes

Sulayman D. Dib-Hajj,[*,†,‡] **Yong Yang,**[§] **and Stephen G. Waxman**[*,†,‡]

*Department of Neurology, Yale University School of Medicine,
New Haven, Connecticut 06510
†Center for Neuroscience and Regeneration Research, Yale University School
of Medicine, New Haven, Connecticut 06510
‡Rehabilitation Research Center, VA Connecticut Healthcare System,
West Haven, Connecticut 06516
§Department of Dermatology, Peking University First Hospital,
Beijing 100034, China

I. Introduction
   A. Tissue distribution and subcellular localization of sodium channels
   B. Discovery and electrophysiological properties of $Na_v1.7$
II. Role of $Na_v1.7$ in Pain Syndromes: Animal Studies
   A. Traumatic injury
   B. Inflammation-induced pain
   C. Painful diabetic neuropathy
III. $Na_v1.7$ and Inherited Pain Syndromes
   A. Inherited erythromelalgia
   B. Paroxysmal extreme pain disorder
   C. $Na_v1.7$-related congenital insensitivity to pain
IV. Conclusions
   Acknowledgments
   References

Advances in Genetics, Vol. 63
Copyright 2008, Elsevier Inc. All rights reserved.

0065-2660/08 $35.00
DOI: 10.1016/S0065-2660(08)01004-3

ABSTRACT

SCN9A, the gene which encodes voltage-gated sodium channel $Na_v1.7$, is located on human chromosome 2 within a cluster of other members of this gene family. $Na_v1.7$ is present at high levels in most peripheral nociceptive neurons in dorsal root ganglion (DRG) and in sympathetic neurons. In addition to its focal tissue-specific expression, $Na_v1.7$ is distinguished by its ability to amplify small depolarizations, thus acting as a threshold channel and modulating excitability. Dominantly inherited gain-of-function mutations in SCN9A have been linked to two familial painful disorders: inherited erythromelalgia (IEM) and paroxysmal extreme pain disorder (PEPD). One set of mutations leads to severe episodes of pain in the feet and hands in patients with IEM, and a different set of mutations causes pain in a perirectal, periocular, and mandibular distribution in patients with PEPD. These mutations allow mutant channels to activate in response to weaker stimuli, or to remain open longer in response to stimulation. The introduction of mutant channels into DRG neurons alters electrogenesis and renders these primary sensory neurons hyperexcitable. Mutant $Na_v1.7$ channels lower the threshold for single action potentials and increase the number of action potentials that neurons fire in response to suprathreshold stiumli. In contrast, recessively inherited loss-of-function mutations in SCN9A, which cause a loss of function of $Na_v1.7$ in patients, lead to indifference to pain with sparing of motor and cognitive abilities. The central role of $Na_v1.7$ in these disorders, and the apparently limited consequences of loss of this channel in humans make it an attractive target for treatment of pain. © 2008, Elsevier Inc.

I. INTRODUCTION

Voltage-gated sodium channels open transiently in response to stimuli that depolarize the plasma membrane, thus allowing the flow of sodium ions down their concentration gradient, and generating an inward current which underlies the generation and conduction of action potential (AP) in excitable cells. Sodium channels are heteromultimers of a large pore-forming α-subunit (will be referred to as channels throughout the chapter) and smaller auxiliary β-subunits (Catterall, 2000). β-subunits and an array of other cytosolic channel partners regulate trafficking and anchoring of the channels at the cell membrane, or modulate biophysical properties of the channels either by direct binding or by inducing posttranslational modifications, for example, phosphorylation (Abriel and Kass, 2005; Cantrell and Catterall, 2001; Hudmon et al., 2008; Liu et al., 2005; Wittmack et al., 2005; Wood et al., 2004).

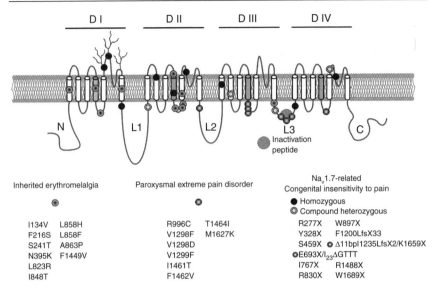

Figure 4.1. Schematic of voltage-gated sodium channel showing locations of the known mutations in Na$_v$1.7-related inherited pain disorders. Inherited erythromelalgia (IEM, ⊛ symbols) and paroxysmal extreme pain disorder (PEPD, ● symbols) mutations are gain-of-function and inherited as a dominant trait. Na$_v$1.7-related congenital indifference to pain (CIP) is caused by loss-of-function mutations which are inherited as a recessive trait. Homozygous Na$_v$1.7-related CIP mutations carry the same nonsense mutation on both alleles of SCN9A (● symbols), whereas two pairs of compound heterozygous mutations (◉ symbols) carry different mutations which produce nonfunctional channels on the two alleles.

Sodium channels are large polypeptides (size range 1700–2000 amino acids) which are organized into four domains (DI–DIV), each consisting of six transmembrane segments that are connected by intra- and extracellular linkers (Fig. 4.1; Catterall, 2000). Nine members of a gene family that encodes sodium channels (Na$_v$1.1–Na$_v$1.9) and several of their cognates have been identified in mammals and lower vertebrates (Catterall *et al.*, 2005; Goldin, 2002; Goldin *et al.*, 2000). Many of these channels produce sodium currents with distinct biophysical and pharmacological properties (Rush *et al.*, 2007). Expression of sodium channels is regulated in spatial and temporal patterns, and individual neurons express multiple sodium channels.

A. Tissue distribution and subcellular localization of sodium channels

Sodium channel expression, even for channels that are encoded by genes within the same cluster on a specific chromosome, is tissue specific across the different species where it has been investigated. Neuronal sodium channels Na$_v$1.1–1.3

and $Na_v1.6-1.9$ can be found within CNS and PNS neurons, whereas $Na_v1.4$ is present in skeletal and $Na_v1.5$ in cardiac myocytes (Catterall *et al.*, 2005); we will not discuss further nonneuronal channels in this chapter. Some of these channels are sorted to specific neuronal compartments which suggest a specialized function in the initiation and conduction of the electrical signal.

SCN1A–3A and *SCN9A*, the genes for sodium channels $Na_v1.1$, $Na_v1.2$, $Na_v1.3$, and $Na_v1.7$, are clustered on human chromosome 2 and are thought to have arisen by gene duplication (Goldin, 2002). Despite their common ancestry and chromosomal location, these channels are expressed in different tissues or at different developmental stages. $Na_v1.1$ is primarily expressed in both CNS (Beckh *et al.*, 1989) and dorsal root ganglion (DRG) neurons (Black *et al.*, 1996). $Na_v1.2$ is primarily a CNS channel which is expressed at low levels during embryogenesis and reaches adult levels in the rat by P14 (Beckh *et al.*, 1989; Felts *et al.*, 1997; Westenbroek *et al.*, 1989). $Na_v1.3$ is the predominant isoform during embryogenesis in CNS and DRG neurons in the rat and its expression is significantly attenuated after birth in most tissues (Beckh *et al.*, 1989; Waxman *et al.*, 1994), but is detected in sympathetic ganglia in adult rodents, at levels comparable to those of other channels (Rush *et al.*, 2006). However, adult human CNS neurons continue to produce higher levels of $Na_v1.3$ compared with their rodent counterparts (Whitaker *et al.*, 2001). $Na_v1.7$ is expressed mainly in sensory and sympathetic neurons in the PNS (Black *et al.*, 1996; Felts *et al.*, 1997; Rush *et al.*, 2006; Sangameswaran *et al.*, 1997; Toledo-Aral *et al.*, 1997), and in myenteric neurons (Sage *et al.*, 2007).

SCN8A, the gene which encodes $Na_v1.6$, is located on human chromosome 12 (Burgess *et al.*, 1995). $Na_v1.6$ is expressed in CNS neurons (Burgess *et al.*, 1995; Schaller *et al.*, 1995; Whitaker *et al.*, 1999) and in DRG neurons (Black *et al.*, 1996; Dietrich *et al.*, 1998) and sympathetic ganglia (Rush *et al.*, 2006). The expression of $Na_v1.6$ in rat CNS neurons increases after birth, reaching adult levels by P14 (Schaller and Caldwell, 2000), whereas it is detected within DRG neurons at E17 and becomes robust by P0 (Chung *et al.*, 2003; Felts *et al.*, 1997).

SCN10A and *SCN11A*, the genes which encode $Na_v1.8$ and $Na_v1.9$, respectively, are located on chromosome 3 (Dib-Hajj *et al.*, 1999; Souslova *et al.*, 1997), in a cluster with *SCN5A*, the gene which encodes the cardiac channel $Na_v1.5$. $Na_v1.8$ and $Na_v1.9$ are sensory neuron-specific channels which are normally expressed within DRG and trigeminal ganglia but not in CNS neurons (Akopian *et al.*, 1996; Dib-Hajj *et al.*, 1998; Sangameswaran *et al.*, 1996). $Na_v1.8$ is present in most peptidergic and nonpeptidergic small DRG neurons (Amaya *et al.*, 2000; Djouhri *et al.*, 2003; Fang *et al.*, 2005; Fjell *et al.*, 1999; Rush *et al.*, 2005; Sleeper *et al.*, 2000), whereas $Na_v1.9$ expression has been reported mainly in small-diameter nonpeptidergic IB4[+] neurons (Amaya *et al.*, 2000; Fang *et al.*, 2002; Fjell *et al.*, 1999; Rush *et al.*, 2005).

These channels are distinguished from the other neuronal channels by their voltage-dependent and kinetic properties and by their resistance to micromolar concentrations of tetrodotoxin (Catterall *et al.*, 2005).

Na$_v$1.1 and Na$_v$1.3 appear to be predominantly localized to somatodendritic compartments of myelinated neurons (Westenbroek *et al.*, 1989; Whitaker *et al.*, 2001). However, Na$_v$1.1 was detected at the initial segment of axons in the inner plexiform layer of the retina (Van Wart *et al.*, 2005), but not in optic nerve axons (Craner *et al.*, 2003); the subcellular distribution of Na$_v$1.1 in adult rat DRG neurons is not well defined. Na$_v$1.2 is present along premyelinated axons and is targeted to immature nodes of Ranvier during myelination of CNS axons (Boiko *et al.*, 2001; Kaplan *et al.*, 2001), but is then restricted to the somatodendritic compartment, stretches of axons that lack myelin, and totally nonmyelinated axons (Boiko *et al.*, 2001; Westenbroek *et al.*, 1989, 1992). Considering the very low levels of Na$_v$1.2 in rodent DRG neurons after birth (Black *et al.*, 1996; Waxman *et al.*, 1994), it is not surprising that this channel is not detected at nodes of Ranvier in peripheral myelinated axons (Schafer *et al.*, 2006). Na$_v$1.6 is the main sodium channel isoform at mature nodes of Ranvier in the CNS (Boiko *et al.*, 2001; Kaplan *et al.*, 2001) and the PNS (Black *et al.*, 2002; Schafer *et al.*, 2006), and is present in the somata and diffusely along substantial lengths of unmyelinated fibers arising from small DRG neurons (Black *et al.*, 2002; Rush *et al.*, 2005).

The three peripheral channels Na$_v$1.7, Na$_v$1.8, and Na$_v$1.9 are primarily localized to the soma of small DRG neurons and along small-diameter myelinated (Aδ) and unmyelinated C-fibers. Na$_v$1.7 has been detected in the somata of DRG (Toledo-Aral *et al.*, 1997), sympathetic neurons (Rush *et al.*, 2006), and myenteric neurons (Sage *et al.*, 2007), and along unmyelinated fibers in the sciatic nerve (Rush *et al.*, 2005) and within the neurite tips of DRG and trigeminal neurons in culture (Dib-Hajj *et al.*, 2007; Toledo-Aral *et al.*, 1997). Na$_v$1.8 is abundant in the somata of small DRG neurons (Amaya *et al.*, 2000; Sleeper *et al.*, 2000), along unmyelinated fibers in the sciatic nerve (Rush *et al.*, 2005) and trigeminal nerve (Black and Waxman, 2002), and is present at some nodes in PNS (Black *et al.*, 2008). Na$_v$1.8 is also abundant within nerve terminals in the skin (Zhao *et al.*, 2008) and cornea (Black and Waxman, 2002). Na$_v$1.9 is abundant in the somata of the majority of small nonpeptidergic DRG neurons (Amaya *et al.*, 2000; Sleeper *et al.*, 2000), and along unmyelinated fibers and nerve endings in the cornea (Black and Waxman, 2002) and in the skin (Dib-Hajj *et al.*, 2002; Rush *et al.*, 2005), and is also present at some nodes along small-diameter myelinated PNS fibers (Fjell *et al.*, 2000).

B. Discovery and electrophysiological properties of Na$_v$1.7

Na$_v$1.7 was first isolated from human medullary thyroid carcinoma (hMTC) cell line, and called hNE-Na (Klugbauer *et al.*, 1995), and from a rabbit Schwann cell library and called NaS (Belcher *et al.*, 1995), and later cloned from rat DRG and called PN1 (Sangameswaran *et al.*, 1997; Toledo-Aral *et al.*, 1997). SCN9A, the gene which encodes $Na_v1.7$, is composed of 26 coding exons spaning 167.3 Mb on chromosome 2 (2q24). Recently, the promoter and 5' noncoding exons which span a 64 Mb have been identified in human SCN9A (Diss *et al.*, 2007). Similar to few other sodium channels, exon 5 (92 bp) is present sequentially in mutually exclusive isoforms, embryonic/neonatal (E5N) and adult (E5A) (Belcher *et al.*, 1995; Raymond *et al.*, 2004). The amino acid sequence of the alternative exon 5 differs by only two residues: E5N (L201; N206) and E5A (V201; D206). Additionally, the utilization of an alternative 5' splice site for intron 12 lengthens exon 11 by 33 bp, leading to an extension of loop 1 (L1) which joins domains 1 and 2 by 11 amino acids (Raymond *et al.*, 2004). The two forms of L1 are designated L1-short (S) and L1-long (L). The abundance of the four isoforms ($Na_v1.7$/NS, $Na_v1.7$/NL, $Na_v1.7$/AS, and $Na_v1.7$/AL) varies during development (Raymond *et al.*, 2004). A similar situation has also been reported for $Na_v1.2$ (Raymond *et al.*, 2004) and $Na_v1.6$ (Dietrich *et al.*, 1998).

$Na_v1.7$ produces a fast activating and inactivating current which is sensitive to nanomolar concentrations of tetrodotoxin (TTX-S) (Klugbauer *et al.*, 1995; Sangameswaran *et al.*, 1997). $Na_v1.7$ is distinguished among TTX-S sodium channels by slow recovery from fast inactivation (slow repriming) (Cummins *et al.*, 1998; Herzog *et al.*, 2003). $Na_v1.7$ undergoes slow closed-state inactivation which permits it to pass a current (ramp current) in response to small slow depolarizations (Cummins *et al.*, 1998; Herzog *et al.*, 2003). Comparison of the biophysical properties of alternative isoforms of $Na_v1.7$ in a heterologous expression system (human embryonic kidney 293 cell line) suggests that the most dramatic difference is slower kinetics of inactivation at negative potentials for isoforms carrying the E5A exon, which is associated with bigger ramp currents, and loss of modulation by 8Br-cAMP for isoforms carrying the longer L1 (Chatelier *et al.*, 2008). The ability of $Na_v1.7$ to respond to ramp stimuli suggests that it might act as a "threshold" channel, amplifying generator potentials, and thus setting the gain in nociceptors where it is coexpressed with $Na_v1.8$ (Cummins *et al.*, 2007; Rush *et al.*, 2007; Waxman and Hains, 2006).

II. ROLE OF $NA_V1.7$ IN PAIN SYNDROMES: ANIMAL STUDIES

A. Traumatic injury

The role of $Na_v1.7$ in traumatic nerve injury in animal models is not very well understood. $Na_v1.7$ knockout studies suggested that this channel is not essential for injury-induced pain (Nassar *et al.*, 2005). There is a reduction in transcript levels of $Na_v1.7$ in DRG neurons following axotomy (Kim *et al.*, 2001, 2002), which is accompanied by the reduction in the slowly repriming TTX-S sodium currents (Cummins *et al.*, 1998). These findings are also consistent with studies on human patients with traumatic injury to their peripheral nerves, which shows that Na$_v$1.7 immunostaining is significantly reduced in injured DRG neurons compared with control (Coward *et al.*, 2001). However, Na$_v$1.7 has been reported to accumulate within neuromas of trunk nerves (Black *et al.*, 2008; Coward *et al.*, 2000; Kretschmer *et al.*, 2002). Recently, Na$_v$1.7 accumulation has been reported in painful neuromas of the lingual nerve, but not in nonpainful neuromas of the same nerve (Bird *et al.*, 2007). Studies on human patients share the common limitation of small numbers of samples and a paucity of control tissue, two deficiencies that stress the need for caution in interpreting these studies.

B. Inflammation-induced pain

The contribution of Na$_v$1.7 to acquired neuropathies is best documented in animal studies of inflammatory pain (Black *et al.*, 2004; Nassar *et al.*, 2004). Peripheral tissue inflammation causes a significant increase in the amplitude of the TTX-S current in DRG neurons (Black *et al.*, 2004), which is accompanied by increases in Na$_v$1.7 transcript and protein levels which are more robust than for Na$_v$1.3, the other TTX-S channel that is upregulated under these conditions (Black *et al.*, 2004; Gould *et al.*, 2004). In agreement with this conclusion, knockdown of Na$_v$1.7 in primary afferents *in vivo* prevents thermal hyperalgesia which is induced by injecting hindpaw of mice with complete Freund's adjuvant (Yeomans *et al.*, 2005). Perhaps, the most convincing evidence for the central role of Na$_v$1.7 in inflammatory pain is the total abrogation of inflammation-induced mechanical and thermal hyperalgesia in mice with DRG-specific knockout of this channel (Nassar *et al.*, 2004).

C. Painful diabetic neuropathy

Streptozotocin-induced diabetic neuropathy (STZ) causes sodium channel dysregulation which is accompanied by tactile allodynia 6 weeks following the injection of the drug (Craner *et al.*, 2002). In this study, dysregulated expression of several sodium channels was evident in diabetic rats 8 weeks after the onset of allodynia with significant increases in levels of Na$_v$1.3, Na$_v$1.6, and Na$_v$1.9, but with no change in Na$_v$1.7 (Craner *et al.*, 2002). However, Na$_v$1.7 protein was reported to increase in another study together with an increase in the TTX-S ramp current in these neurons (Hong *et al.*, 2004). The increase in ramp current is consistent with an upregulation of either Na$_v$1.3 or Na$_v$1.7 (or both) channels since these two channels have been shown to produce a robust ramp current in DRG neurons and in heterologous expression systems (Cummins *et al.*, 1998, 2001; Herzog *et al.*, 2003). The increase in ramp currents in diabetic neurons may contribute to a lowered threshold for firing in these neurons.

III. NA$_V$1.7 AND INHERITED PAIN SYNDROMES

Three inherited pain disorders have been linked to mutations in Na$_v$1.7 (Dib-Hajj *et al.*, 2007; Drenth and Waxman, 2007). Mutations in sodium channel Na$_v$1.7 have been identified in patients suffering from inherited erythromelalgia (IEM) (Dib-Hajj *et al.*, 2007) and paroxysmal extreme pain disorder (PEPD) (Fertleman *et al.*, 2006). Recently, individuals with congenital and complete loss of Na$_v$1.7 have been reported to be "indifferent" to pain while showing no apparent deficits in other sensory modalities (Cox *et al.*, 2006). The mutations that result in painful symptoms are invariably dominantly inherited gain-of-function substitutions, whereas the insensitivity to pain results from recessively inherited loss-of-function mutations (Dib-Hajj *et al.*, 2007; Drenth and Waxman, 2007). Heterozygous carriers of the loss-of-function mutations show normal pain behavior which indicates no haploinsufficiency (Cox *et al.*, 2006).

A. Inherited erythromelalgia

1. Historical background, types, symptoms, and management of IEM

Erythromelalgia (*erythros* = red; *melos* = extremities; *algos* = pain) was first recognized as a medical entity in 1878 by Silas Weir Mitchell (Mitchell, 1878). The inherited form is also called primary erythromelalgia or erythermalgia (van Genderen *et al.*, 1993). Secondary, noninherited forms of this disorder are associated with myeloproliferative diseases such as polycythemia vera, cutaneous vasculitis, systemic lupus erythematosus, hypertension, rheumatoid arthritis, and diabetes mellitus, and have been reported as possible complications of treatment with drugs, for example, verapamil, nifedipine, nicardipine, pergolide, and bromocriptine (Drenth and Michiels, 1994; Drenth *et al.*, 1994). This chapter will focus on inherited forms of erythromelalgia (IEM) including simplex cases in which new mutations appear *de novo* in patients whose parents are asymptomatic.

  Symptoms of early-onset (as early as 1-year old) IEM are characterized by episodes of burning pain triggered by mild warmth or exercise, together with erythema and mild swelling in the hands and feet, and sometimes in the ears or face (Drenth and Michiels, 1992; Hisama *et al.*, 2006; Novella *et al.*, 2007; van Genderen *et al.*, 1993). The frequency and severity of pain episodes increase with age, with each episode lasting minutes to hours. Despite the presence of Na$_v$1.7 in sympathetic neurons, IEM patients do not report global autonomic abnormalities such as orthostatic hypotension or gastrointestinal symptoms. Neurological examinations of these patients are normal, consistent with unremarkable MRI brain scans and sensory and motor nerve conduction studies.

Pharmacotherapy has been largely ineffective, and partial relief of symptoms comes from cooling the affected extremities (Dib-Hajj et al., 2007; Drenth and Waxman, 2007).

2. Linkage of IEM to chromosome 2 and SCN9A

Erythromelalgia symptoms in a large kindred from the USA were shown to be transmitted in a classical Mendelian autosomal-dominant pattern (Finley et al., 1992), but the underlying gene remained unknown for more than a decade. Linkage analysis of this large American kindred (Fig. 4.2A) defined a 7.94-cM

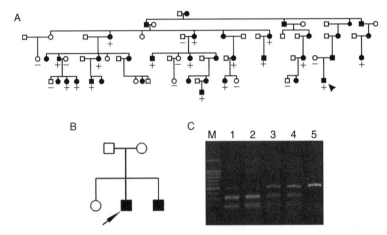

Figure 4.2. Pedigrees of familial and sporadic cases of IEM. (A) F1449V mutation in Na$_v$1.7 in a familial case of IEM from the USA. Circles denote females; squares denote males. The proband is shown by an arrowhead. Blackened symbols indicate subjects affected with erythromelalgia. (+) denotes subjects heterozygous for the F1449V mutation; (−) denotes subjects without the mutation. Reproduced with permission from Dib-Hajj et al. (2005). (B) L858F mutation in Na$_v$1.7 in a sporadic case of IEM from China. Family pedigree showing affected brothers with sporadic erythermalgia together with unaffected sister and mother and an asymptomatic father. Filled symbols denote clinically affected individuals. Arrow denotes proband. (C) Asymptomatic father shows genetic mosaicism at C2572 position. The C2572T mutation destroys the recognition site for the restriction enzyme BstPI. Digestion of wild-type exon 15 amplicon (470 bp) yields two fragments of 300 and 170 bp. The mother's sample shows complete digestion of the amplicon into two fragments (lane 2). Digestion of amplicons from the affected siblings shows the presence of three bands (lanes 3 and 4) consistent with the presence of wild-type and mutant alleles. Digestion of the father's amplicon shows the presence of three bands consistent with the presence of the mutant and wild-type alleles. Signal intensity is quantitated assuming a 1:1 ratio of intact/cut in the children's samples (lanes 3 and 4); the corresponding ratio in the father's sample (lane 1) is 0.15:1. "M" is the 100-bp marker. Lane 5 contains uncut DNA. Reproduced with permission from Han et al. (2006).

segment on human chromosome 2, and haplotype analysis on several other families with erythromelalgia showed a shared haplotype on chromosome 2q31–32 (Drenth *et al.*, 2001), but the responsible gene was not identified. The localization of the erythromelalgia locus to chromosome 2q was then independently confirmed and narrowed to a 5.98-cM interval in a three-generation family from China by Yang *et al.* (2004). Based on the analysis of unique clinical features of IEM, evocable, symmetrical and episodic pain in the extremities, they inferred that it might be a channelopathy. A candidate gene, sodium channel SCN9A which encodes the peripheral sodium channel Na$_v$1.7, is present within this genomic interval, and sequence analysis showed that it carries a different missense mutation in each of the two Chinese families (Yang *et al.*, 2004). The large American kindred (Fig. 4.2A) was later found to carry a third missense mutation (Dib-Hajj *et al.*, 2005). The confirmation of disease-causing potential of these mutations was provided by demonstration of significant effects on gating properties of mutant Na$_v$1.7 channels, which made these channels easier to open compared with wild-type channels (Cummins *et al.*, 2004) and made DRG neurons hyperexcitable (Dib-Hajj *et al.*, 2005) as will be discussed in more details later.

3. Molecular genetics of IEM

Ten missense mutations in SCN9A (Table 4.1) have been identified in familial and sporadic cases of IEM (Dib-Hajj *et al.*, 2005; Drenth *et al.*, 2005; Han *et al.*, 2006; Harty *et al.*, 2006; Lee *et al.*, 2007; Michiels *et al.*, 2005; Takahashi *et al.*, 2007; Yang *et al.*, 2004). In some cases (Table 4.1), the same mutation has been described in sporadic and familial cases, indicating that all of the SCN9A-related IEM mutations are heritable. Several mutations such as I848T and L858F have been reported in several unrelated cases, suggesting a hot spot for mutations in this part of the channel gene (Zhang *et al.*, 2007). The mutations span the channel through domain III, with no mutations in domain IV, but with half the mutations concentrated in DII (Table 4.1).

A few cases of IEM have reported two substitutions in the SCN9A sequence of the proband, for example, P610T/L858H from a Canadian family (Drenth *et al.*, 2005) and A863P/R1150W in a sporadic case from the USA (Harty *et al.*, 2006). The contribution of P610T to the disease phenotype is not clear at this time since L858F was the only mutation reported in a Chinese family (Fig. 4.2B) and this substitution alone is sufficient to alter the biophysical properties of mutant Na$_v$1.7 (Han *et al.*, 2006). Recently, analysis of the Canadian family has shown that P610T segregates independently of L858F and, importantly, is present in one unaffected sibling in this family and in 10/210 ethnically matched control chromosomes (Samuels *et al.*, 2008). Similarly,

Table 4.1. Na$_v$1.7 Mutations from Patients with IEM

Patient ethnicity (nationality)	Nucleotide change	Exon	Mutation	Channel region	No. of cases	References
Chinese (Taiwan)	c.406A>G	3	I136V	DI/S1	1	(Lee et al., 2007)
Canadian	c.647T>C	5	F216S	DI/S4	1	(Drenth et al., 2005)
Flemish	c.721T>A	6	S241T	DI/S4–S5	1	(Michiels et al., 2005)
Dutch	c.1185C>A	9	N395K	DI/S6	2	(Drenth et al., 2005; Zhang et al., 2007)
Japanese	c.2468T>G	14	L823R	DII/S3–S4	1	(Takahashi et al., 2007)
Chinese/French	c.2543T>C	15	I848T	DII/S4–S5	5	(Drenth et al., 2005; Yang et al., 2004; Zhang et al., 2007)
Chinese/Canadian	c.2572C>T	15	L858F	DII/S4–S5	2	(Drenth et al., 2005; Han et al., 2006)
Chinese	c.2573T>A	15	L858H	DII/S4–S5	2	(Yang et al., 2004; Zhang et al., 2007)
USA	c.2587G>C	15	A863P	DII/S5	1	(Harty et al., 2006)
USA	c.4345T>G	23	F1449V	DIII/S6	1	(Dib-Hajj et al., 2005)

A863P alone is sufficient to alter the biophysical properties of mutant Na$_v$1.7 and render DRG neurons hyperexcitable (Harty et al., 2006). However, it remains to be seen if the compound genotype modulates the severity of the symptoms compared with the causative mutation alone.

Along with several other de novo mutations (Yang et al., 2004; Zhang et al., 2007), a genetic mosaicism case was found in China (Han et al., 2006). An asymptomatic couple bore three children, two of whom were affected with IEM (Fig. 4.2B). After sequence analysis of SCN9A, a L858F mutation was detected in both affected siblings, but not in the healthy mother and sister. Although the father had no clinical manifestation, further restriction enzyme and sequencing analysis demonstrated that 13% of the SCN9A DNA in his blood carried the L858F mutation (Fig. 4.2C), and thus 26% of his blood cells carry the mutant allele. The lack of symptoms could be a result of the low number of affected cells, or, alternatively, the mutation may not be present in peripheral neurons (Han et al., 2006).

4. Mutation-induced changes in biophysical properties of Na$_v$1.7

Whole-cell voltage-clamp studies of mutant Na$_v$1.7 channels in mammalian cells have shown that all of the IEM mutations cause a lowering of the threshold for activation of Na$_v$1.7, and many slow deactivation, which is the transition of the channel from open to closed state, and increase the ramp response of the channel (Fig. 4.3) (Cheng *et al.*, 2008; Choi *et al.*, 2006; Cummins *et al.*, 2004; Dib-Hajj *et al.*, 2005; Han *et al.*, 2006; Harty *et al.*, 2006; Lampert *et al.*, 2006; Sheets *et al.*, 2007). Each of these changes can contribute to DRG neuron hyperexcitability. These studies have not investigated the effect of natural variants of the channel on the behavior of the mutant channels, for example, the contribution of P610T or R1150W to the overall phenotype of the channels that carry these variants and a causative mutation.

The reported ineffectiveness of pharmacotherapy in the treatment of IEM can be linked to the behavior of the mutant channel in the presence of these drugs. It should not be surprising that mexiletine and lidocaine do not ameliorate pain symptoms in patients carrying the N395K mutation (Drenth *et al.*, 2005), for example, because this mutation reduces the binding affinity of these drugs to the mutant channel (Sheets *et al.*, 2007). Carbamazepine, another sodium channel blocker that is effective in the treatment of PEPD (see next section) but not in IEM, does not reduce the hyperpolarized shift in the voltage dependence of activation of I848T mutant $Na_v1.7$ channel (Fertleman *et al.*, 2006). However, it is important to consider that other IEM mutations may enhance the response of the mutant channel to these drugs, and thus might be effective in treating the symptoms. This possibility will have to be empirically determined in the future.

5. Mutation-induced changes in DRG neuron firing

Whole-cell current-clamp studies have thus far been carried out on DRG neurons transfected with three mutant $Na_v1.7$ channels and have demonstrated neuronal hyperexcitability compared with cells transfected with wild-type $Na_v1.7$ (Dib-Hajj *et al.*, 2005; Harty *et al.*, 2006; Rush *et al.*, 2006). These studies have shown that mutant $Na_v1.7$ channels lower the threshold for single action potential and increase the firing frequency of neurons to graded stimuli (Fig. 4.4), both features of hyperexcitable nociceptors. While the three mutations share the common feature of a hyperpolarized shift in voltage dependence of activation, the magnitude of the shift and changes in other gating properties are not similar, which suggests a dependence on the shift of activation as the primary cause of neuronal hyperexcitability. Indeed, computer simulations are consistent with the idea that while changes in other channel gating properties can modulate the response of the neuron, the lowering of the threshold for mutant $Na_v1.7$ activation is most important in rendering DRG neurons hyperexcitable (Sheets *et al.*, 2007). The mutant $Na_v1.7$-induced hyperexcitability of nociceptive DRG neurons could explain the burning pain that is associated with IEM.

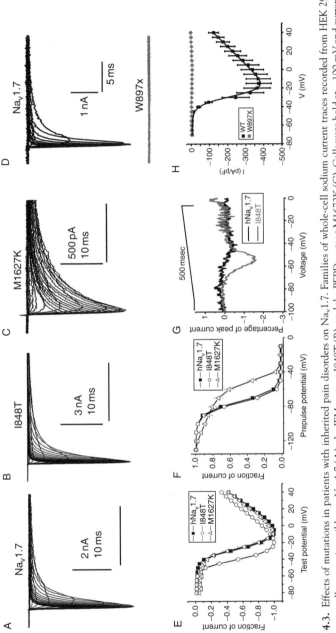

Figure 4.3. Effects of mutations in patients with inherited pain disorders on $Na_v1.7$. Families of whole-cell sodium current traces recorded from HEK 293 cells expressing wild-type $Na_v1.7$ (A), the IEM mutant I848T (B), and the PEPD mutant M1672K (C). Cells were held at -100 mV and currents were elicited with 50-ms test depolarizations to potentials ranging from -80 to 40 mV. (D) *top trace, black:* Family of whole-cell sodium current traces from wild-type $Na_v1.7$ which is transfected into HEK 293 cells. Cells were held at -100 mV followed by depolarization steps of 50 ms to voltages between -70 and $+40$ mV in 5-mV increments at 0.5 Hz. (D) *bottom trace, grey:* $Na_v1.7$ carrying the homozygous null $Na_v1.7$-related CIP W897X mutation produces nonfunctional channels. (E) The I848T erythromelalgia mutation shifts the normalized peak current–voltage relationship in a hyperpolarizing direction compared with wild-type and M1672K $Na_v1.7$ channels. (F) The M1672K PEPD mutation shifts the

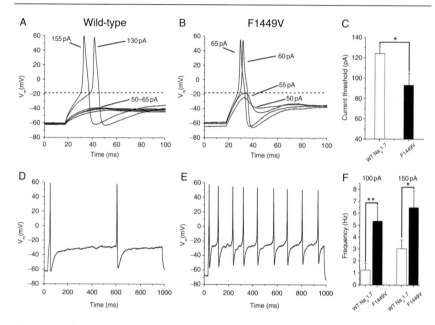

Figure 4.4. Effect of the F1449V mutation on electrogenesis in small DRG neurons. (A) Representative traces from a small (<30 μm) DRG neuron expressing wild-type Na$_v$1.7, showing subthreshold responses to 50–65 pA current injections and subsequent all-or-none action potentials evoked by injections of 130 pA (current threshold for this neuron) and 155 pA. (B) In contrast, in a cell expressing the F1449V erythromelalgia mutation, action potentials were evoked by a 60-pA current injection, demonstrating a lower current threshold for action potential generation. The voltage for takeoff of the all-or-none action potential (dotted line) was similar for the neurons in panels (A) and (B). (C) There is a significant (*p < 0.05) reduction in current threshold in cells expressing F1449V compared with cells expressing wild-type Na$_v$1.7. (D) Shows the firing of a neuron expressing wild-type Na$_v$1.7 (same neuron as in (A)), which responded to a 950-ms stimulation of 150 pA with two action potentials. In contrast, (E) shows that, in a neuron expressing the mutant channel F1449V (same cell as in (B)), an identical 150-pA depolarizing stimulus evoked high-frequency firing. (F) There is a significant increase in the frequency of firing in response to 100 and 150 pA stimuli (950 ms) following expression of F1449V in comparison with wild-type Na$_v$1.7. Modified with permission from Dib-Hajj *et al.* (2005).

voltage dependence of steady-state fast inactivation in a depolarizing direction compared with wild-type and I848T Na$_v$1.7 channels. (G) The I848T erythromelalgia mutation enhances ramp currents. Representative ramp currents elicited with 500-ms ramp depolarizations from −100 to 0 mV are shown. (H) Normalized current–voltage relationship of peak currents obtained from HEK 293 cells which were transfected with wild-type or Na$_v$1.7-related CIP mutant channel W897X. None of Na$_v$1.7-related CIP mutant Na$_v$1.7 constructs (S458X or I767X) produces functional channels. Reproduced with permission from Cummins *et al.* (2004) and Dib-Hajj *et al.* (2008). (D) and (H) modified with permission from Cox *et al.* (2006).

B. Paroxysmal extreme pain disorder

1. Historical background, symptoms, and management of PEPD

PEPD is a pain disorder that was first identified as a distinct clinical entity in 1959 (Hayden and Grossman, 1959) and initially coined familial rectal pain because of its prominent feature of painful bowel movement (Dugan, 1972). Recently a consortium of patients and treating physicians agreed to rename familial rectal pain as PEPD (Fertleman and Ferrie, 2006). Lifelong PEPD symptoms start typically at birth and are triggered by bowel movement, falling on the buttocks, or probing of the genital area, and are accompanied by nonepileptic seizures and immediately followed by flushing of the lower limbs, in a uni- or bilateral fashion (Fertleman et al., 2007). The frequency of rectal pain episodes decreases with age, with painful episodes becoming more prominent in ocular and mandibular regions, sometimes triggered by cold or irritants or food (Fertleman et al., 2007). Typically, patients with PEPD respond favorably to treatment with carbamazepine which reduces the frequency and intensity of pain episodes (Fertleman et al., 2006).

2. Linkage of PEPD to chromosome 2 and SCN9A

PEPD symptoms in a large kindred were shown to be transmitted in a classical Mendelian autosomal-dominant pattern (Dugan, 1972). A genome-wide linkage analysis on a different large kindred from the UK narrowed down the locus to a 16-cM region on human chromosome 2, which includes the cluster of voltage-gated sodium channel genes (Fertleman et al., 2006). The candidate gene, sodium channel SCN9A, was screened and missense mutations were found in coding exons in 8/13 cases (Fertleman et al., 2006). The absence of SCN9A mutations in five cases with PEPD is consistent with linkage analysis of a family from the USA which produced a negative LOD score, suggesting the presence of targets other than SCN9A (Fertleman et al., 2006). It is possible, however, that mutations in noncoding exons or in introns may increase SCN9A expression and contribute to the mutant phenotype.

3. Molecular genetics of PEPD

Eight mutations (Fig. 4.1 and Table 4.2) have been found in Na$_v$1.7 from eight families with PEPD and segregated with the disease in affected families (Fertleman et al., 2006). A second-generation affected patient carries a second mutation (R996C) in addition to the one (V1298D) he inherited from his affected father (Fertleman et al., 2006). It is not known if the two mutations are in cis or trans configuration. The symptoms in the patient with the compound mutations appear to be more severe compared with those of his father (Fertleman et al., 2006).

Table 4.2. Na$_v$1.7 Mutations from Patients with PEPD

Patient ethnicity (nationality)	Nucleotide change	Exon	Mutation	Channel region	No. of cases	References
UK	c.2986C>T	16	R996C	L2	1	(Fertleman *et al.*, 2006)
UK	c.3940G>T	21	V1298F	DIII/S4–S5	1	(Fertleman *et al.*, 2006)
French	c.3941T>A	21	V1298D	DIII/S4–S5	1	(Fertleman *et al.*, 2006)
UK	c.3943G>T	21	V1299F	DIII/S4–S5	1	(Fertleman *et al.*, 2006)
UK	c.4382T>C	24	I1461T	L3 IFM	1	(Fertleman *et al.*, 2006)
UK	c.4384T>G	24	F1462V	L3 IFM	1	(Fertleman *et al.*, 2006)
UK	c.4391C>T	24	T1463I	L3 IFM	1	(Fertleman *et al.*, 2006)
UK/French	c.4879T>A	26	M1627K	DIV/S4–S5	2	(Dib-Hajj *et al.*, 2008; Fertleman *et al.*, 2006)

Except for one mutation (R99C) which is located in L2, the other mutations are concentrated in domain III (three mutations within the N-terminus of the S4–S5 linker), L3 (three mutations that affect the IFMT inactivation peptide), and one mutation in domain IV (middle of S4–S5 linker) (Fig. 4.1 and Table 4.2). These regions of the channel have been implicated in fast inactivation of sodium channels (Catterall, 2000).

4. Mutation-induced changes in biophysical properties of Na$_v$1.7

The location of PEPD mutations (Fig. 4.1 and Table 4.2) suggested an effect on fast inactivation of mutant Na$_v$1.7 channel. Whole-cell patch-clamp studies of three mutations (I1461T, T1461I, and M1627K) were initially carried out in mammalian HEK 293 cell line (Fertleman *et al.*, 2006). As predicted, the I1461T and T1461I mutations, which affect the fast-inactivation tetrapeptide IFMT, cause a depolarizing shift and incomplete steady-state fast inactivation, and have been reported to produce a substantial persistent current, but with no effect on channel activation (Fertleman *et al.*, 2006). Recently, another group studied the mutations V1298F, V1299F, and I1461T and confirmed the depolarizing shift in steady-state inactivation but reported much smaller persistent current, compared with what was previously reported (Jarecki *et al.*, 2008). In contrast, M1762K shifts the voltage dependence of steady-state fast inactivation, but inactivation is complete (Fertleman *et al.*, 2006) (Fig. 4.3). The voltage dependence of slow inactivation, a state that is likely to arise during repetitive stimulation of neurons, or the application of a strong stimulus for a relatively prolonged time (seconds instead of milliseconds), is impaired by the PEPD mutations in the voltage range around the resting membrane potential of neurons (Jarecki *et al.*, 2008). The depolarizing shift in the voltage dependence of fast and slow inactivation permits more channels to be available during depolarization of the neuron, and is predicted to contribute to enhanced window currents that allow more sodium current to flow through the mutant channel, which could be linked to repetitive firing in hyperexcitable DRG neurons (Catterall and Yu, 2006). In fact, one PEPD mutation that has been studied by current-clamp to date, M1627K, has been observed to increase the firing frequency of DRG neurons in response to a graded suprathreshold stimuli (Dib-Hajj et al., 2008).

PEPD patients are remarkable in responding to treatment with carbamazepine (Bednarek et al., 2005; Fertleman et al., 2006; Schubert and Cracco, 1992). Carbamazepine is a use-dependent inhibitor of sodium channels that preferentially binds to inactivated channels and enhances inactivation of all tested sodium channels (Priest and Kaczorowski, 2007). Carbamazepine has been shown to enhance fast inactivation of the peak current, and inhibit the persistent sodium current that is produced by the PEPD I1461T and T1464I mutant channels (Fertleman et al., 2006). Thus, carbamazepine may achieve efficacy by enhancing inactivation and stabilizing the inactivated state of mutant channels, consistent with its ameliorative effect in PEPD patients.

C. Na$_v$1.7-related congenital insensitivity to pain

1. Historical background and symptoms

Congenital insensitivity (or indifference) to pain (CIP) is a rare syndrome which has been linked to many target genes, and patients with some forms of CIP may show other deficits, for example, anhydrosis, in addition to sensing, perceiving, and reacting to painful stimuli (Nagasako et al., 2003; Verhoeven et al., 2006). A more pure form of CIP, affecting only pain sensation and olfaction, has recently been linked to loss-of-function mutations in Na$_v$1.7 (Cox et al., 2006; Goldberg et al., 2007). Patients with Na$_v$1.7-related CIP present with a history of not ever experiencing any pain even after burns, bone fractures, lip- and tongue-biting, and they do not experience visceral pain (Ahmad et al., 2007; Cox et al., 2006; Goldberg et al., 2007). Additionally, patients with Na$_v$1.7-related CIP do not show apparent sympathetic dysfunction (Cox et al., 2006; Goldberg et al., 2007) and have a normal axon reflex response to histamine (Goldberg et al., 2007). Interestingly, however, patients with Na$_v$1.7-related CIP show a deficit in the sense of smell (Goldberg et al., 2007). Nerve biopsy, conduction studies, and brain MRI of patients were all within normal range (Cox et al., 2006).

2. Linkage to chromosome 2 and SCN9A

Genome-wide linkage analysis was first reported for three families from Pakistan which narrowed down the locus for these CIP cases to a 11.7 Mb on human chromosome 2q24, which includes a cluster of sodium channel genes

(Cox *et al.*, 2006). Candidate gene analysis focused on *SCN9A* as the best candidate, and sequence analysis identified distinct homozygous nonsense mutations in the three families which segregated with the disease phenotype in each family (Cox *et al.*, 2006). Similar studies on nine additional families from Europe, USA, and Argentina linked nonsense mutations in *SCN9A* to these cases (Goldberg *et al.*, 2007). Parents of these patients are asymptomatic, consistent with a recessive pattern of inheritance, and indicating that single allele null genotype does not lead to haploinsufficiency (Ahmad *et al.*, 2007; Cox *et al.*, 2006; Goldberg *et al.*, 2007).

3. Molecular genetics of $Na_v1.7$-related CIP

Sequence analysis of *SCN9A* from 12 families with $Na_v1.7$-related CIP identified 11 nonsense mutations (Fig. 4.1 and Table 4.3), with one mutation (R277X) appearing in two unrelated cases (Goldberg *et al.*, 2007). Most of the mutations (9) are homozygous single nucleotide substitutions which result in nonsense mutations that truncate the protein. Two compound heterozygous mutations which include nonsense mutation on one allele and a deletion mutation of coding, or intronic sequences (suggests splicing defect), on the other allele have also been found in the remaining two cases (Fig. 4.1 and Table 4.3). $Na_v1.7$-related CIP patients are ethnically diverse and geographically isolated, suggesting that these mutations arose independently. While most of the $Na_v1.7$ null mutations reported to date are identified in offsprings of consanguineous marriages (Cox *et al.*, 2006; Goldberg *et al.*, 2007), several mutations have been identified in smaller kindreds (CIP-05, CIP-26, CIP-32, CIP-33, and CIP-102) with nonconsanguineous marriages (P. Goldberg, personal communication).

4. Molecular pathophysiology of $Na_v1.7$-related CIP

Homozygous and compound null mutations in *SCN9A* (Fig. 4.1 and Table 4.3) are predicted to truncate the channel protein, resulting in loss-of-function mutations in $Na_v1.7$ and the complete loss of $Na_v1.7$ current in all of the neurons in which this channel is expressed. Indeed, null mutants (Y328X, S459X, I767X, and W897X) of $Na_v1.7$ channel do not produce functional channels in mammalian cell lines (Fig. 4.3) (Ahmad *et al.*, 2007; Cox *et al.*, 2006). There is no evidence that the truncated channel protein accumulates or that it interferes with other sodium channels that may be present in the same cell (Ahmad *et al.*, 2007).

Table 4.3. Mutations from Patients with $Na_v1.7$-Related CIP

Patient ethnicity (nationality)	Nucleotide change	Exon	Mutation	Channel region	No. of cases	References
Swiss	c.829C>T	6	R277X	DI/S5–S6	2	(Goldberg et al., 2007)
Canadian	c.984C>A	8	Y328X	DI/S5–S6	1	(Goldberg et al., 2007)
Pakistani	c.1376C>G	10	S459X	L1	1	(Cox et al., 2006)
USA	c.2076-2077insT	13	E693X/I23ΔGTTT	L1	1	(Goldberg et al., 2007)
	c.4366-7_10delGTTT	123		L3		
Pakistani	c.2298delT	13	I767X	DII/S2	1	(Cox et al., 2006)
French	c.2488C>T	15	R830X	DII/S4	1	(Goldberg et al., 2007)
Pakistani	c.2691G>A	15	W897X	DII/S5–S6	1	(Cox et al., 2006)
Italian	c.3600delT	18	F1200fsX33	DIII/S1–S2	1	(Goldberg et al., 2007)
UK	c.3703_3713del	19	Δ11bpI1235fsX2/K1659X	DIII/S3	1	(Goldberg et al., 2007)
	c.4975A>T	26		DIV/S5–S6		
USA	c.4462C>T	24	R1488X	L3	1	(Goldberg et al., 2007)
Argentinian	c.5067G>A	26	W1689X	DIV/SS2	1	(Goldberg et al., 2007)

Two cases of $Na_v1.7$-related CIP are compound heterozygous (*italic type*) because they carry two independent null mutations on the two SCN9A alleles.

IV. CONCLUSIONS

We have presented evidence in this chapter to support the conclusion that inherited $Na_v1.7$ channelopathies underlie painful symptoms in two distinct clinical entities: IEM and PEPD. Dysregulated expression of $Na_v1.7$ under a variety of pathological conditions induced by trauma or inflammation, and metabolic disorders, for example, diabetes, may contribute to acquired pain in these disorders. While the role of $Na_v1.7$ in amplifying weak stimuli might be consistent with the effect of gain-of-function mutations or increased channel density in acquired painful disorders, it is less clear why the absence of $Na_v1.7$ totally does prevent the pain signal in the CIP patients from being transmitted along the peripheral–central pain axis. Irrespectively, the key role of $Na_v1.7$ in neuropathic pain conditions and the absence of cardiac, cognitive, or motor deficits in $Na_v1.7$-related CIP make this channel an attractive target for the development of new pharmacological agents for the treatment of pain.

Acknowledgments

This work was supported by the Medical Research Service and Rehabilitation Research Service, Department of Veterans Affairs and by a grant from the Erythromelalgia Association (SDH and SGW); National High Technology Research and Development Programme of China (Grant 2002BA711A07) and National Natural Science Foundation (Grant 30400168) (YY). The Center for Neuroscience and Regeneration Research is a Collaboration of the Paralyzed Veterans of America, and the United Spinal Association with Yale University.

References

Abriel, H., and Kass, R. S. (2005). Regulation of the voltage-gated cardiac sodium channel Na(v)1.5 by interacting proteins. *Trends Cardiovasc. Med.* **15,** 35–40.

Ahmad, S., Dahllund, L., Eriksson, A. B., Hellgren, D., Karlsson, U., Lund, P. E., Meijer, I. A., Meury, L., Mills, T., Moody, A., Morinville, A., Morten, J., *et al.* (2007). A stop codon mutation in SCN9A causes lack of pain sensation. *Hum. Mol. Genet.* **16,** 2114–2121.

Akopian, A. N., Sivilotti, L., and Wood, J. N. (1996). A tetrodotoxin-resistant voltage-gated sodium channel expressed by sensory neurons. *Nature* **379,** 257–262.

Amaya, F., Decosterd, I., Samad, T. A., Plumpton, C., Tate, S., Mannion, R. J., Costigan, M., and Woolf, C. J. (2000). Diversity of expression of the sensory neuron-specific TTX-resistant voltage-gated sodium ion channels SNS and SNS2. *Mol. Cell. Neurosci.* **15,** 331–342.

Beckh, S., Noda, M., Lubbert, H., and Numa, S. (1989). Differential regulation of three sodium channel messenger RNAs in the rat central nervous system during development. *EMBO J.* **8,** 3611–3616.

Bednarek, N., Arbues, A. S., Motte, J., Sabouraud, P., Plouin, P., and Morville, P. (2005). Familial rectal pain: A familial autonomic disorder as a cause of paroxysmal attacks in the newborn baby. *Epileptic Disord.* **7,** 360–362.

Belcher, S. M., Zerillo, C. A., Levenson, R., Ritchie, J. M., and Howe, J. R. (1995). Cloning of a sodium channel alpha subunit from rabbit Schwann cells. *Proc. Natl Acad. Sci. USA* **92,** 11034–11038.

Bird, E. V., Robinson, P. P., and Boissonade, F. M. (2007). Na(v)1.7 sodium channel expression in human lingual nerve neuromas. *Arch. Oral Biol.* **52**(5), 494–502.

Black, J. A., and Waxman, S. G. (2002). Molecular identities of two tetrodotoxin-resistant sodium channels in corneal axons. *Exp. Eye Res.* **75**, 193–199.

Black, J. A., Dib-Hajj, S., McNabola, K., Jeste, S., Rizzo, M. A., Kocsis, J. D., and Waxman, S. G. (1996). Spinal sensory neurons express multiple sodium channel alpha-subunit mRNAs. *Mol. Brain Res.* **43**, 117–131.

Black, J. A., Renganathan, M., and Waxman, S. G. (2002). Sodium channel Na(v)1.6 is expressed along nonmyelinated axons and it contributes to conduction. *Mol. Brain Res.* **105**, 19–28.

Black, J. A., Liu, S., Tanaka, M., Cummins, T. R., and Waxman, S. G. (2004). Changes in the expression of tetrodotoxin-sensitive sodium channels within dorsal root ganglia neurons in inflammatory pain. *Pain* **108**, 237–247.

Black, J., Nikolajsen, L., Kroner, K., Jensen, T., and Waxman, S. (2008). Multiple sodium channel isoforms and MAP kinases are present in painful human neuromas. *Ann. Neurol.* In Press.

Boiko, T., Rasband, M. N., Levinson, S. R., Caldwell, J. H., Mandel, G., Trimmer, J. S., and Matthews, G. (2001). Compact myelin dictates the differential targeting of two sodium channel isoforms in the same axon. *Neuron* **30**, 91–104.

Burgess, D. L., Kohrman, D. C., Galt, J., Plummer, N. W., Jones, J. M., Spear, B., and Meisler, M. H. (1995). Mutation of a new sodium channel gene, Scn8a, in the mouse mutant 'motor endplate disease'. *Nat. Genet.* **10**, 461–465.

Cantrell, A. R., and Catterall, W. A. (2001). Neuromodulation of Na$^+$ channels: An unexpected form of cellular plasticity. *Nat. Rev. Neurosci.* **2**, 397–407.

Catterall, W. A. (2000). From ionic currents to molecular mechanisms: The structure and function of voltage-gated sodium channels. *Neuron* **26**, 13–25.

Catterall, W. A., and Yu, F. H. (2006). Painful channels. *Neuron* **52**, 743–744.

Catterall, W. A., Goldin, A. L., and Waxman, S. G. (2005). International Union of Pharmacology. XLVII. Nomenclature and structure–function relationships of voltage-gated sodium channels. *Pharmacol. Rev.* **57**, 397–409.

Chatelier, A., Dahllund, L., Eriksson, A., Krupp, J., and Chahine, M. (2008). Biophysical properties of human Nav1.7 splice variants and their regulation by protein kinase A. *J. Neurophysiol* **99**, 2241–2250.

Cheng, X., Dib-Hajj, S. D., Tyrrell, L., and Waxman, S. G. (2008). Mutation I136V alters electrophysiological properties of the NaV1.7 channel in a family with onset of erythromelalgia in the second decade. *Mol. Pain* **4**, 1.

Choi, J. S., Dib-Hajj, S. D., and Waxman, S. G. (2006). Inherited erythermalgia. Limb pain from an S4 charge-neutral Na channelopathy. *Neurology* **67**, 1563–1567.

Chung, J. M., Dib-Hajj, S. D., and Lawson, S. N. (2003). Sodium channel subtypes and neuropathic pain. In "Proceedings of the 10th Congress in Pain Research and Management" (J. O. Dostrovsky, D. B. Carr, and M. Koltzenberg, eds.), Vol. 24, pp. 99–114. IASP Press, Seatle.

Coward, K., Plumpton, C., Facer, P., Birch, R., Carlstedt, T., Tate, S., Bountra, C., and Anand, P. (2000). Immunolocalization of SNS/PN3 and NaN/SNS2 sodium channels in human pain states. *Pain* **85**, 41–50.

Coward, K., Aitken, A., Powell, A., Plumpton, C., Birch, R., Tate, S., Bountra, C., and Anand, P. (2001). Plasticity of TTX-sensitive sodium channels PN1 and brain III in injured human nerves. *Neuroreport* **12**, 495–500.

Cox, J. J., Reimann, F., Nicholas, A. K., Thornton, G., Roberts, E., Springell, K., Karbani, G., Jafri, H., Mannan, J., Raashid, Y., Al-Gazali, L., Hamamy, H., et al. (2006). An SCN9A channelopathy causes congenital inability to experience pain. *Nature* **444**, 894–898.

Craner, M. J., Klein, J. P., Renganathan, M., Black, J. A., and Waxman, S. G. (2002). Changes of sodium channel expression in experimental painful diabetic neuropathy. *Ann. Neurol.* **52**, 786–792.

Craner, M. J., Lo, A. C., Black, J. A., and Waxman, S. G. (2003). Abnormal sodium channel distribution in optic nerve axons in a model of inflammatory demyelination. *Brain* **126**, 1552–1561.

Cummins, T. R., Howe, J. R., and Waxman, S. G. (1998). Slow closed-state inactivation: A novel mechanism underlying ramp currents in cells expressing the hNE/PN1 sodium channel. *J. Neurosci.* **18**, 9607–9619.

Cummins, T. R., Aglieco, F., Renganathan, M., Herzog, R. I., Dib-Hajj, S. D., and Waxman, S. G. (2001). Nav1.3 sodium channels: Rapid repriming and slow closed-state inactivation display quantitative differences after expression in a mammalian cell line and in spinal sensory neurons. *J. Neurosci.* **21**, 5952–5961.

Cummins, T. R., Dib-Hajj, S. D., and Waxman, S. G. (2004). Electrophysiological properties of mutant Nav1.7 sodium channels in a painful inherited neuropathy. *J. Neurosci.* **24**, 8232–8236.

Cummins, T. R., Sheets, P. L., and Waxman, S. G. (2007). The roles of sodium channels in nociception: Implications for mechanisms of pain. *Pain* **131**, 243–257.

Dib-Hajj, S. D., Tyrrell, L., Black, J. A., and Waxman, S. G. (1998). NaN, a novel voltage-gated Na channel, is expressed preferentially in peripheral sensory neurons and down-regulated after axotomy. *Proc. Natl Acad. Sci. USA* **95**, 8963–8968.

Dib-Hajj, S. D., Tyrrell, L., Escayg, A., Wood, P. M., Meisler, M. H., and Waxman, S. G. (1999). Coding sequence, genomic organization, and conserved chromosomal localization of the mouse gene Scn11a encoding the sodium channel NaN. *Genomics* **59**, 309–318.

Dib-Hajj, S., Black, J. A., Cummins, T. R., and Waxman, S. G. (2002). NaN/Nav1.9: A sodium channel with unique properties. *Trends Neurosci.* **25**, 253–259.

Dib-Hajj, S. D., Rush, A. M., Cummins, T. R., Hisama, F. M., Novella, S., Tyrrell, L., Marshall, L., and Waxman, S. G. (2005). Gain-of-function mutation in Nav1.7 in familial erythromelalgia induces bursting of sensory neurons. *Brain* **128**, 1847–1854.

Dib-Hajj, S. D., Cummins, T. R., Black, J. A., and Waxman, S. G. (2007). From genes to pain: Na$_v$1.7 and human pain disorders. *Trends Neurosci.* **30**, 555–563.

Dib-Hajj, S., Estacion, M., Jarecki, B., Tyrrell, L., Fischer, T., Lawden, M., Cummins, T., and Waxman, S. (2008). Paroxysmal extreme pain disorder M1627K mutation in human Nav1.7 renders DRG neurons hyperexcitable. *Mol. Pain* **4**, 37.

Dietrich, P. S., McGivern, J. G., Delgado, S. G., Koch, B. D., Eglen, R. M., Hunter, J. C., and Sangameswaran, L. (1998). Functional analysis of a voltage-gated sodium channel and its splice variant from rat dorsal root ganglia. *J. Neurochem.* **70**, 2262–2272.

Diss, J. K., Calissano, M., Gascoyne, D., Djamgoz, M. B., and Latchman, D. S. (2007). Identification and characterization of the promoter region of the Nav1.7 voltage-gated sodium channel gene (SCN9A). *Mol. Cell. Neurosci.* **37**, 537–47.

Djouhri, L., Fang, X., Okuse, K., Wood, J. N., Berry, C. M., and Lawson, S. (2003). The TTX-resistant sodium channel Nav1.8 (SNS/PN3): Expression and correlation with membrane properties in rat nociceptive primary afferent neurons. *J. Physiol. (Lond.)* **550**, 739–752.

Drenth, J. P., and Michiels, J. J. (1992). Clinical characteristics and pathophysiology of erythromelalgia and erythermalgia. *Am. J. Med.* **93**, 111–114.

Drenth, J. P., and Michiels, J. J. (1994). Erythromelalgia and erythermalgia: Diagnostic differentiation. *Int. J. Dermatol.* **33**, 393–397.

Drenth, J. P., and Waxman, S. G. (2007). Mutations in sodium-channel gene SCN9A cause a spectrum of human genetic pain disorders. *J. Clin. Invest.* **117**, 3603–3609.

Drenth, J. P., van Genderen, P. J., and Michiels, J. J. (1994). Thrombocythemic erythromelalgia, primary erythermalgia, and secondary erythermalgia: Three distinct clinicopathologic entities. *Angiology* **45**, 451–453.

Drenth, J. P., Finley, W. H., Breedveld, G. J., Testers, L., Michiels, J. J., Guillet, G., Taieb, A., Kirby, R. L., and Heutink, P. (2001). The primary erythermalgia-susceptibility gene is located on chromosome 2q31–32. *Am. J. Hum. Genet.* **68**, 1277–1282.

Drenth, J. P., Te Morsche, R. H., Guillet, G., Taieb, A., Kirby, R. L., and Jansen, J. B. (2005). SCN9A mutations define primary erythermalgia as a neuropathic disorder of voltage gated sodium channels. *J. Invest. Dermatol.* **124**, 1333–1338.

Dugan, R. E. (1972). Familial rectal pain. *Lancet* **1**, 854.

Fang, X., Djouhri, L., Black, J. A., Dib-Hajj, S. D., Waxman, S. G., and Lawson, S. N. (2002). The presence and role of the tetrodotoxin-resistant sodium channel Na(v)1.9 (NaN) in nociceptive primary afferent neurons. *J. Neurosci.* **22**, 7425–7433.

Fang, X., Djouhri, L., McMullan, S., Berry, C., Okuse, K., Waxman, S. G., and Lawson, S. N. (2005). trkA is expressed in nociceptive neurons and influences electrophysiological properties via Nav1.8 expression in rapidly conducting nociceptors. *J. Neurosci.* **25**, 4868–4878.

Felts, P. A., Yokoyama, S., Dib-Hajj, S., Black, J. A., and Waxman, S. G. (1997). Sodium channel alpha-subunit mRNAs I, II, III, NaG, Na6 and HNE (PN1)—Different expression patterns in developing rat nervous system. *Mol. Brain Res.* **45**, 71–82.

Fertleman, C. R., and Ferrie, C. D. (2006). What's in a name—Familial rectal pain syndrome becomes paroxysmal extreme pain disorder. *J. Neurol. Neurosurg. Psychiatry* **77**, 1294–1295.

Fertleman, C. R., Baker, M. D., Parker, K. A., Moffatt, S., Elmslie, F. V., Abrahamsen, B., Ostman, J., Klugbauer, N., Wood, J. N., Gardiner, R. M., and Rees, M. (2006). SCN9A mutations in paroxysmal extreme pain disorder: Allelic variants underlie distinct channel defects and phenotypes. *Neuron* **52**, 767–774.

Fertleman, C. R., Ferrie, C. D., Aicardi, J., Bednarek, N. A., Eeg-Olofsson, O., Elmslie, F. V., Griesemer, D. A., Goutieres, F., Kirkpatrick, M., Malmros, I. N., Pollitzer, M., Rossiter, M., et al. (2007). Paroxysmal extreme pain disorder (previously familial rectal pain syndrome). *Neurology* **69**, 586–595.

Finley, W. H., Lindsey, J. R., Jr, Fine, J. D., Dixon, G. A., and Burbank, M. K. (1992). Autosomal dominant erythromelalgia. *Am. J. Med. Genet.* **42**, 310–315.

Fjell, J., Cummins, T. R., Dib-Hajj, S. D., Fried, K., Black, J. A., and Waxman, S. G. (1999). Differential role of GDNF and NGF in the maintenance of two TTX-resistant sodium channels in adult DRG neurons. *Mol. Brain Res.* **67**, 267–282.

Fjell, J., Hjelmstrom, P., Hormuzdiar, W., Milenkovic, M., Aglieco, F., Tyrrell, L., Dib-Hajj, S., Waxman, S. G., and Black, J. A. (2000). Localization of the tetrodotoxin-resistant sodium channel NaN in nociceptors. *Neuroreport* **11**, 199–202.

Goldberg, Y., Macfarlane, J., Macdonald, M., Thompson, J., Dube, M. P., Mattice, M., Fraser, R., Young, C., Hossain, S., Pape, T., Payne, B., Radomski, C., et al. (2007). Loss-of-function mutations in the Na$_v$1.7 gene underlie congenital indifference to pain in multiple human populations. *Clin. Genet.* **71**, 311–319.

Goldin, A. L. (2002). Evolution of voltage-gated Na(+) channels. *J. Exp. Biol.* **205**, 575–584.

Goldin, A. L., Barchi, R. L., Caldwell, J. H., Hofmann, F., Howe, J. R., Hunter, J. C., Kallen, R. G., Mandel, G., Meisler, M. H., Netter, Y. B., Noda, M., Tamkun, M. M., et al. (2000). Nomenclature of voltage-gated sodium channels. *Neuron* **28**, 365–368.

Gould, H. J., III, England, J. D., Soignier, R. D., Nolan, P., Minor, L. D., Liu, Z. P., Levinson, S. R., and Paul, D. (2004). Ibuprofen blocks changes in Na(v) 1.7 and 1.8 sodium channels associated with complete Freund's adjuvant-induced inflammation in rat. *J. Pain* **5**, 270–280.

Han, C., Rush, A. M., Dib-Hajj, S. D., Li, S., Xu, Z., Wang, Y., Tyrrell, L., Wang, X., Yang, Y., and Waxman, S. G. (2006). Sporadic onset of erythermalgia: A gain-of-function mutation in Nav1.7. *Ann. Neurol.* **59**, 553–558.

Harty, T. P., Dib-Hajj, S. D., Tyrrell, L., Blackman, R., Hisama, F. M., Rose, J. B., and Waxman, S. G. (2006). Na$_V$1.7 mutant A863P in erythromelalgia: Effects of altered activation and steady-state inactivation on excitability of nociceptive dorsal root ganglion neurons. *J. Neurosci.* **26,** 12566–12575.

Hayden, R., and Grossman, M. (1959). Rectal, ocular, and submaxillary pain; a familial autonomic disorder related to proctalgia fugax: Report of a family. *AMA J. Dis. Child.* **97,** 479–482.

Herzog, R. I., Cummins, T. R., Ghassemi, F., Dib-Hajj, S. D., and Waxman, S. G. (2003). Distinct repriming and closed-state inactivation kinetics of Nav1.6 and Nav1.7 sodium channels in mouse spinal sensory neurons. *J. Physiol. (Lond.)* **551,** 741–750.

Hisama, F., Dib-Hajj, S., and Waxman, S. (2006). SCN9A-related inherited erythromelalgia. In GeneReviews at GeneTests: Medical genetics information resource. http://www.genetests.org

Hong, S., Morrow, T. J., Paulson, P. E., Isom, L. L., and Wiley, J. W. (2004). Early painful diabetic neuropathy is associated with differential changes in tetrodotoxin-sensitive and -resistant sodium channels in dorsal root ganglion neurons in the rat. *J. Biol. Chem.* **279,** 29341–29350.

Hudmon, A., Choi, J. S., Tyrrell, L., Black, J. A., Rush, A. M., Waxman, S. G., and Dib-Hajj, S. D. (2008). Phosphorylation of sodium channel Na$_V$1.8 by p38 mitogen-activated protein kinase increases current density in dorsal root ganglion neurons. *J. Neurosci.* **28,** 3190–3201.

Jarecki, B. W., Sheets, P. L., Jackson Ii, J. O., and Cummins, T. R. (2008). Paroxysmal extreme pain disorder mutations within the D3/S4–S5 linker of Nav1.7 cause moderate destabilization of fast-inactivation. *J. Physiol. (Lond.)* **586**(Pt. 17), 4137–4153.

Kaplan, M. R., Cho, M., Ullian, E. M., Isom, L. L., Levinson, S. R., and Barres, B. A. (2001). Differential control of clustering of the sodium channels Na(v)1.2 and Na(v)1.6 at developing CNS nodes of Ranvier. *Neuron* **30,** 105–119.

Kim, C. H., Oh, Y., Chung, J. M., and Chung, K. (2001). The changes in expression of three subtypes of TTX sensitive sodium channels in sensory neurons after spinal nerve ligation. *Mol. Brain Res.* **95,** 153–161.

Kim, C. H., Oh, Y., Chung, J. M., and Chung, K. (2002). Changes in three subtypes of tetrodotoxin sensitive sodium channel expression in the axotomized dorsal root ganglion in the rat. *Neurosci. Lett.* **323,** 125–128.

Klugbauer, N., Lacinova, L., Flockerzi, V., and Hofmann, F. (1995). Structure and functional expression of a new member of the tetrodotoxin-sensitive voltage-activated sodium channel family from human neuroendocrine cells. *EMBO J.* **14,** 1084–1090.

Kretschmer, T., Happel, L. T., England, J. D., Nguyen, D. H., Tiel, R. L., Beuerman, R. W., and Kline, D. G. (2002). Clinical article accumulation of PN1 and PN3 sodium channels in painful human neuroma-evidence from immunocytochemistry. *Acta Neurochir. (Wien)* **144,** 803–810.

Lampert, A., Dib-Hajj, S. D., Tyrrell, L., and Waxman, S. G. (2006). Size matters: Erythromelalgia mutation S241T in Nav1.7 alters channel gating. *J. Biol. Chem.* **281,** 36029–36035.

Lee, M. J., Yu, H. S., Hsieh, S. T., Stephenson, D. A., Lu, C. J., and Yang, C. C. (2007). Characterization of a familial case with primary erythromelalgia from Taiwan. *J. Neurol.* **254,** 210–214.

Liu, C., Cummins, T. R., Tyrrell, L., Black, J. A., Waxman, S. G., and Dib-Hajj, S. D. (2005). CAP-1A is a novel linker that binds clathrin and the voltage-gated sodium channel Na(v)1.8. *Mol. Cell. Neurosci.* **28,** 636–649.

Michiels, J. J., te Morsche, R. H., Jansen, J. B., and Drenth, J. P. (2005). Autosomal dominant erythermalgia associated with a novel mutation in the voltage-gated sodium channel alpha subunit Nav1.7. *Arch. Neurol.* **62,** 1587–1590.

Mitchell, S. W. (1878). On a rare vaso-motor neurosis of the extremities, and on the maladies with which it may be confounded. *Am. J. Med. Sci.* **76,** 17–36.

Nagasako, E. M., Oaklander, A. L., and Dworkin, R. H. (2003). Congenital insensitivity to pain: An update. *Pain* **101,** 213–219.

Nassar, M. A., Stirling, L. C., Forlani, G., Baker, M. D., Matthews, E. A., Dickenson, A. H., and Wood, J. N. (2004). Nociceptor-specific gene deletion reveals a major role for Nav1.7 (PN1) in acute and inflammatory pain. *Proc. Natl Acad. Sci. USA* 101, 12706–12711.

Nassar, M. A., Levato, A., Stirling, C., and Wood, J. N. (2005). Neuropathic pain develops normally in mice lacking both Nav1.7 and Nav1.8. *Mol. Pain* 1, 24.

Novella, S. P., Hisama, F. M., Dib-Hajj, S. D., and Waxman, S. G. (2007). A case of inherited erythromelalgia. *Nat. Clin. Pract. Neurol.* 3, 229–234.

Priest, B. T., and Kaczorowski, G. J. (2007). Subtype-selective sodium channel blockers promise a new era of pain research. *Proc. Natl Acad. Sci. USA* 104, 8205–8206.

Raymond, C. K., Castle, J., Garrett-Engele, P., Armour, C. D., Kan, Z., Tsinoremas, N., and Johnson, J. M. (2004). Expression of alternatively spliced sodium channel alpha-subnit genes: Unique splicing patterns are observed in dorsal root ganglia. *J. Biol. Chem.* 279, 46234–46241.

Rush, A. M., Craner, M. J., Kageyama, T., Dib-Hajj, S. D., Waxman, S. G., and Ranscht, B. (2005). Contactin regulates the current density and axonal expression of tetrodotoxin-resistant but not tetrodotoxin-sensitive sodium channels in DRG neurons. *Eur. J. Neurosci.* 22, 39–49.

Rush, A. M., Dib-Hajj, S. D., Liu, S., Cummins, T. R., Black, J. A., and Waxman, S. G. (2006). A single sodium channel mutation produces hyper- or hypoexcitability in different types of neurons. *Proc. Natl Acad. Sci. USA* 103, 8245–8250.

Rush, A. M., Cummins, T. R., and Waxman, S. G. (2007). Multiple sodium channels and their roles in electrogenesis within dorsal root ganglion neurons. *J. Physiol. (Lond.)* 579(Pt. 1), 1–14.

Sage, D., Salin, P., Alcaraz, G., Castets, F., Giraud, P., Crest, M., Mazet, B., and Clerc, N. (2007). Na$_v$1.7 and Na$_v$1.3 are the only tetrodotoxin-sensitive sodium channels expressed by the adult guinea pig enteric nervous system. *J. Comp. Neurol.* 504, 363–378.

Samuels, M. E., Te Morsche, R. H., Lynch, M. E., and Drenth, J. P. (2008). Compound heterozygosity in sodium channel Nav1.7 in a family with hereditary erythermalgia. *Mol. Pain* 4, 21.

Sangameswaran, L., Delgado, S. G., Fish, L. M., Koch, B. D., Jakeman, L. B., Stewart, G. R., Sze, P., Hunter, J. C., Eglen, R. M., and Herman, R. C. (1996). Structure and function of a novel voltage-gated, tetrodotoxin-resistant sodium channel specific to sensory neurons. *J. Biol. Chem.* 271, 5953–5956.

Sangameswaran, L., Fish, L. M., Koch, B. D., Rabert, D. K., Delgado, S. G., Ilnicka, M., Jakeman, L. B., Novakovic, S., Wong, K., Sze, P., Tzoumaka, E., Stewart, G. R., et al. (1997). A novel tetrodotoxin-sensitive, voltage-gated sodium channel expressed in rat and human dorsal root ganglia. *J. Biol. Chem.* 272, 14805–14809.

Schafer, D. P., Custer, A. W., Shrager, P., and Rasband, M. N. (2006). Early events in node of Ranvier formation during myelination and remyelination in the PNS. *Neuron Glia Biol.* 2, 69–79.

Schaller, K. L., and Caldwell, J. H. (2000). Developmental and regional expression of sodium channel isoform NaCh6 in the rat central nervous system. *J. Comp. Neurol.* 420, 84–97.

Schaller, K. L., Krzemien, D. M., Yarowsky, P. J., Krueger, B. K., and Caldwell, J. H. (1995). A novel, abundant sodium channel expressed in neurons and glia. *J. Neurosci.* 15, 3231–3242.

Schubert, R., and Cracco, J. B. (1992). Familial rectal pain: A type of reflex epilepsy? *Ann. Neurol.* 32, 824–826.

Sheets, P. L., Jackson Ii, J. O., Waxman, S. G., Dib-Hajj, S., and Cummins, T. R. (2007). A Nav1.7 channel mutation associated with hereditary erythromelalgia contributes to neuronal hyperexcitability and displays reduced lidocaine sensitivity. *J. Physiol. (Lond.)* 581, 1019–1031.

Sleeper, A. A., Cummins, T. R., Dib-Hajj, S. D., Hormuzdiar, W., Tyrrell, L., Waxman, S. G., and Black, J. A. (2000). Changes in expression of two tetrodotoxin-resistant sodium channels and their currents in dorsal root ganglion neurons after sciatic nerve injury but not rhizotomy. *J. Neurosci.* 20, 7279–7289.

Souslova, V. A., Fox, M., Wood, J. N., and Akopian, A. N. (1997). Cloning and characterization of a mouse sensory neuron tetrodotoxin-resistant voltage-gated sodium channel gene, Scn10a. *Genomics* 41, 201–209.

Takahashi, K., Saitoh, M., Hoshino, H., Mimaki, M., Yokoyama, Y., Takamizawa, M., Mizuguchi, M., Lin, Z. M., Yang, Y., and Igarashi, T. (2007). A case of primary erythermalgia, wintry hypothermia and encephalopathy. *Neuropediatrics* **38**, 157–159.

Toledo-Aral, J. J., Moss, B. L., He, Z. J., Koszowski, A. G., Whisenand, T., Levinson, S. R., Wolf, J. J., Silossantiago, I., Halegoua, S., and Mandel, G. (1997). Identification of PN1, a predominant voltage-dependent sodium channel expressed principally in peripheral neurons. *Proc. Natl Acad. Sci. USA* **94**, 1527–1532.

van Genderen, P. J., Michiels, J. J., and Drenth, J. P. (1993). Hereditary erythermalgia and acquired erythromelalgia. *Am. J. Med. Genet.* **45**, 530–532.

Van Wart, A., Boiko, T., Trimmer, J. S., and Matthews, G. (2005). Novel clustering of sodium channel Na(v)1.1 with ankyrin-G and neurofascin at discrete sites in the inner plexiform layer of the retina. *Mol. Cell. Neurosci.* **28**, 661–673.

Verhoeven, K., Timmerman, V., Mauko, B., Pieber, T. R., De Jonghe, P., and Auer-Grumbach, M. (2006). Recent advances in hereditary sensory and autonomic neuropathies. *Curr. Opin. Neurol.* **19**, 474–480.

Waxman, S. G., and Hains, B. C. (2006). Fire and phantoms after spinal cord injury: Na$^+$ channels and central pain. *Trends Neurosci.* **29**, 207–215.

Waxman, S. G., Kocsis, J. D., and Black, J. A. (1994). Type III sodium channel mRNA is expressed in embryonic but not adult spinal sensory neurons, and is reexpressed following axotomy. *J. Neurophysiol.* **72**, 466–470.

Westenbroek, R. E., Merrick, D. K., and Catterall, W. A. (1989). Differential subcellular localization of the RI and RII Na$^+$ channel subtypes in central neurons. *Neuron* **3**, 695–704.

Westenbroek, R. E., Noebels, J. L., and Catterall, W. A. (1992). Elevated expression of type II Na+ channels in hypomyelinated axons of shiverer mouse brain. *J. Neurosci.* **12**, 2259–2267.

Whitaker, W., Faull, R., Waldvogel, H., Plumpton, C., Burbidge, S., Emson, P., and Clare, J. (1999). Localization of the type VI voltage-gated sodium channel protein in human CNS. *Neuroreport* **10**, 3703–3709.

Whitaker, W. R., Faull, R. L., Waldvogel, H. J., Plumpton, C. J., Emson, P. C., and Clare, J. J. (2001). Comparative distribution of voltage-gated sodium channel proteins in human brain. *Mol. Brain Res.* **88**, 37–53.

Wittmack, E. K., Rush, A. M., Hudmon, A., Waxman, S. G., and Dib-Hajj, S. D. (2005). Voltage-gated sodium channel Nav1.6 is modulated by p38 mitogen-activated protein kinase. *J. Neurosci.* **25**, 6621–6630.

Wood, J. N., Boorman, J. P., Okuse, K., and Baker, M. D. (2004). Voltage-gated sodium channels and pain pathways. *J. Neurobiol.* **61**, 55–71.

Yang, Y., Wang, Y., Li, S., Xu, Z., Li, H., Ma, L., Fan, J., Bu, D., Liu, B., Fan, Z., Wu, G., Jin, J., Ding, B., Zhu, X., and Shen, Y. (2004). Mutations in SCN9A, encoding a sodium channel alpha subunit, in patients with primary erythermalgia. *J. Med. Genet.* **41**, 171–174.

Yeomans, D. C., Levinson, S. R., Peters, M. C., Koszowski, A. G., Tzabazis, A. Z., Gilly, W. F., and Wilson, S. P. (2005). Decrease in inflammatory hyperalgesia by Herpes vector-mediated knockdown of Na(v)1.7 sodium channels in primary afferents. *Hum. Gene Ther.* **16**, 271–277.

Zhang, L. L., Lin, Z. M., Ma, Z. H., Xu, Z., Yang, Y. L., and Yang, Y. (2007). Mutation hotspots of SCN9A in primary erythermalgia. *Br. J. Dermatol.* **156**, 767–769.

Zhao, P., Barr, T. P., Hou, Q., Dib-Hajj, S. D., Black, J. A., Albrecht, P. J., Petersen, K., Eisenberg, E., Wymer, J. P., Rice, F. L., and Waxman, S. G. (2008). Voltage-gated sodium channel expression in rat and human epidermal keratinocytes: Evidence for a role in pain. *Pain* **139**, 90–105.

Part II

INTERNAL DISEASES

5

Channelopathies of Transepithelial Transport and Vesicular Function

Christian A. Hübner* and Thomas J. Jentsch†

*Department of Clinical Chemistry, University Hospital of the Friedrich-Schiller-Universität, Erlanger Allee 101, D-07747 Jena, Germany
†FMP (Leibniz-Institut für Molekulare Pharmakologie) and MDC (Max-Delbrück-Centrum für Molekulare Medizin), Robert-Rössle-Strasse 10, D-13125 Berlin, Germany

  I. Introduction
 II. Disorders
     A. CFTR and cystic fibrosis
     B. Channelopathies affecting renal function
     C. Hearing loss and channelopathies
     D. Mucolipidosis type IV
     E. Osteopetrosis and lysosomal storage disease
III. Concluding Remarks
     References

ABSTRACT

The transport of ions across cellular membranes is crucial for various functions, including the transport of salt and water across epithelia, the control of electrical excitability of muscle and nerve, and the regulation of cell volume or the acidification and ionic homeostasis of intracellular organelles. This review will focus on ion channel diseases (channelopathies) that are due to impaired transepithelial transport or due to mutations in vesicular ion transporters or channels. When needed for a more comprehensive understanding of organ function, also other ion transport diseases will be mentioned. © 2008, Elsevier Inc.

Advances in Genetics, Vol. 63
Copyright 2008, Elsevier Inc. All rights reserved.

0065-2660/08 $35.00
DOI: 10.1016/S0065-2660(08)01005-5

I. INTRODUCTION

Ion channels are pores for the passive diffusion of ions across biological membranes that can open or shut due to conformational changes to permit an ion flow of more than 100,000 ions per second per molecule. This current depends on electrochemical gradients that are established by the interplay between cotransporters, pumps, and constitutively open ion channels. Ion channels can be classified according to their functional characteristics or according to their molecular structure. Many ion channels are assembled from several identical or homologous subunits, or may include other, nonhomologous proteins. α-subunits usually refer to subunits that are involved in the formation of the ion translocation pathway (pore), whereas β-subunits are often not directly pore-forming but may be important for the assembly, stability, regulation, or targeting of the ion channel complex.

Functional characteristics of ion channels include ion selectivity and gating properties. Selectivity to certain ions (e.g., Na^+, K^+, Ca^{2+}, Cl^-) is determined by the molecular structure of the pore and is unlikely to be changed by associated proteins. Ion channels can open and close their pore, a process described as "gating." Some channels are opened by the binding of a particular substance like neurotransmitters (outside of the cell) or cytoplasmatic messenger molecules (inside of the cell), whereas others are gated by changes of the voltage across the membrane (voltage-gated ion channels) or mechanical stimuli. Gap junctions constitute a separate class of channels that connect the intracellular compartments of adjacent cells and are permeable not only to ions, but also to larger molecules such as cAMP.

Ion channels serve many functions apart from electrical signal transduction: chemical signal transduction (Ca^{2+} as a second messenger), transepithelial transport, regulation of cytoplasmatic or vesicular ion concentration and pH, and regulation of cell volume. Therefore, ion channel diseases are not confined to excitable cells (Table 5.1). Cystic fibrosis is one of the best-known examples. This disease arises from the failure of Cl^- to pass the cAMP-regulated cystic fibrosis transmembrane regulator (CFTR). This results in an accumulation of mucus in pancreatic ducts and in airways, entailing a life-threatening impairment of lung function.

Ion channel diseases arise in a number of different ways. Mutations may result in either gain or loss of function. X-linked disorders, as well as autosomal-recessive, and autosomal-dominant disorders are known. A dominant disease can arise because a 50% reduction of the amount of protein is insufficient (haploinsufficiency), or by a dominant-negative effect in which a mutant subunit encoded by one allele impairs the function of multimeric channel complexes that also contain normal subunits encoded by the WT allele. Gain-of-function mutations may lead to a dominant phenotype if it leads to a "toxic" effect. The list of human

Table 5.1. Channelopathies Related to Epithelial Transport and Vesicular Function

Protein	Ions	Gene	Subunit	Disease
		Epithelial transport		
Kidney				
Polycystin-2	Cations	PKD2	α	Autosomal-dominant polycystic kidney disease
ENaCα	Na^+	SCNN1A	α	Pseudohypoaldosteronism type 1
ENaCβ	Na^+	SCNEB	β	Pseudohypoaldosteronism type 1 Pseudoaldosteronism (Liddle syndrome)
ENaCγ	Na^+	SCNN1G	γ	Pseudohypoaldosteronism type 1 Pseudoaldosteronism (Liddle syndrome)
TRPP2	Na^+	ADPK2	α	Polycystic kidney disease
TRPM6	Na^+	TRPM6	α	Hypomagnesaemia
ClC-Kb	Cl^-	CLCNKB	α	Bartter syndrome type III
Barttin	Cl^-	BSND	β	Bartter syndrome type IV with deafness
Kir1.1/ROMK	K^+	KCNJ1	α	Bartter syndrome
AQP0	H_2O	AQP0/MP26	α	Cataract
AQP1	H_2O	AQP1	α	Urinary concentration defect
AQP2	H_2O	AQP2	α	Diabetes insipidus
TRPC6	Ca^{2+}	TRPC6	α	Focal segmental glomerulosclerosis
Ear				
ClC-Ka	Cl^-	CLCNKA	α	Bartter syndrome type IV with deafness
ClC-Kb	Cl^-	CLCNKB	α	When mutated together
Barttin	Cl^-	BSND	β	Bartter syndrome type IV with deafness
KCNQ1	K^+	KCNQ1	α	Autosomal-dominant long QT with deafness
KCNE1	K^+	KCNE1	α	Autosomal-dominant long QT with deafness
Cx26		GJB2		Deafness (autosomal dominant and recessive)
Cx30		GJB4		Autosomal-dominant deafness
Cx31		GJB3		Autosomal-dominant deafness
Lung				
CFTR	Cl^-	CFTR	α	Cystic fibrosis congenital absence of the vas deferens (CAVD)
		Vesicular function		
TRPML1	$Ca^{2+}, H^+?$	MCOLN1	α	Mucolipidosis
ClC-5	Cl^-/H^+	CLCN5	α	Dent's disease
ClC-7	Cl^-/H^+	CLCN7	α	Osteopetrosis (autosomal recessive or autosomal dominant) (rec. can be associated with lysosomal storage disease)
Ostm1	Cl^-/H^+	OSTM1	β	Autosomal-recessive osteopetrosis with lysosomal storage

diseases, which have been shown to be associated with defects in various types of ion channels, has considerably grown during the past years (Table 5.1). The current review will focus on recently identified ion channel diseases and on channelopathies related to epithelial transport and vesicular function.

II. DISORDERS

A. CFTR and cystic fibrosis

Cystic fibrosis (CF) was recognized as a discrete disease entity in 1938. It is the most common life-limiting autosomal-recessive disorder in the Caucasian population, the heterozygote frequency of disease-related CFTR alleles being approximately one in 20. The disease incidence is roughly 1:3000 live births among Caucasians (Rosenstein and Cutting, 1998) and occurs with lower frequency in other ethnic groups. It affects the epithelia in several organs and results in a complex multisystem disease that includes lung, exocrine pancreas, intestine, male genital tract, hepatobiliary system, and exocrine sweat glands. It is characterized by viscous mucous secretions that plug smaller airways and secretory ducts. This promotes recurrent inflammation and infection that finally result in end-stage lung disease with bronchiectasis and lung fibrosis. Although irrelevant for the disease outcome, CF is also associated with a high salt concentration in the sweat, which can be used as a diagnostic criterion. The clinical manifestations range from early meconium ileus in 10–20% of affected newborns, childhood death because of progressive lung destruction to recurrent sinusitis and bronchitis. More than 95% of affected males are infertile due to congenital agenesis of the vas deferens (CAVD) and azoospermia. Infertility also occurs in men without pulmonary or gastrointestinal manifestations of cystic fibrosis and should hence be considered in the differential diagnosis of male infertility. With the new techniques of assisted reproductive techniques, the number of live births in men with CF is increasing (Popli and Stewart, 2007). In the classical clinical presentation, recurrent pulmonary infections during childhood with gradually progressive obstructive lung disease are disease determining. In the majority of patients, this is accompanied by exocrine pancreatic insufficiency that can lead to intestinal malabsorption. Because of the significant therapeutical progress in the past years, the overall median survival now approaches the fifth decade of life (Popli and Stewart, 2007; Strausbaugh and Davis, 2007).

It was a major breakthrough of molecular genetics when the CFTR gene on chromosome 7q31 was identified in 1989 by positional cloning based on linkage analysis in affected individuals (Riordan et al., 1989). The CFTR gene has 27 exons and encodes a 1480 amino acid protein with a predicted molecular mass of 168 kDa. CFTR belongs to a gene family encoding ATP-binding cassette (ABC) transporters. These transporters, as exemplified by the multidrug resistance protein (mdr or P-glycoprotein), generally transport ("pump") substances under the expenditure of energy supplied by ATP hydrolysis. Despite CFTR's architectural similarity with ABC transporters, it is not an active pump but mediates a highly regulated, passive anion conductance. ATP binding and hydrolysis at the two nucleotide-binding domains (NBDs) (Fig. 5.1) of CFTR are involved in opening and closing its pore (Gadsby et al., 2006), which is still poorly defined. Like other members of the ABC transporter family, CFTR consists of two homologous repeats. Each of these includes six transmembrane spans followed by a highly conserved NBD. CFTR can be phosphorylated by protein kinase A at several residues at a characteristic regulatory "R-domain" that is important for its activation by cAMP.

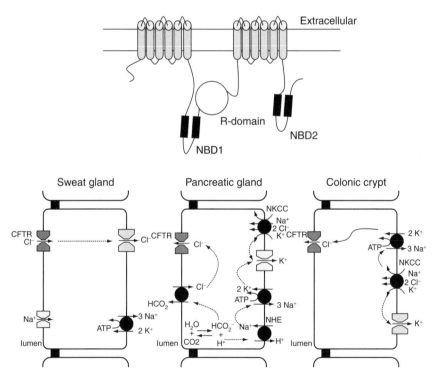

Figure 5.1. CFTR topology and its role in epithelial transport. *Top panel:* Proposed topology of CFTR. *Lower panel:* Suggested functions of CFTR in epithelial cells of sweat glands, pancreatic ducts, and colonic crypts.

The diagnosis of cystic fibrosis is established in patients with typical phenotypic features of the disorder and evidence of impaired CFTR function. This can be either repeated sweat chloride values below 60 mEq/l after pilocarpine iontophoresis or typical transepithelial nasal potential difference measurements or two disease-causing mutations in the *CFTR* gene. In almost all individuals with the clinical presentation of CF, mutations in both alleles can be detected. However, the *CFTR* mutation detection rate varies by test method and ethnic background. As the *CFTR* gene is rather large, usually only the most frequent mutations in a given population are being screened in genetic testing. Various commercially available diagnostic panels exist that are optimized for different populations. Although more than 1000 different mutations have been described so far, a single codon deletion, ΔF508, is by far the most common, with approximately 90% of patients having at least one ΔF508 allele. Phenylalanine-508 is thought to contribute to the binding of the first nucleotide-binding domain (NBD1) to a loop of the transmembrane part of CFTR (Serohijos *et al.*, 2008). Its deletion may interfere with the correct folding of the first NBD and affects gating of the few channel molecules that can make it to the plasma membrane at low temperatures.

The failure of airway epithelia to adequately hydrate mucus on their surfaces is a leading feature in CF. According to the hydration hypothesis (Guggino, 1999), this results from a dysbalance between Cl^--dependent fluid secretion mediated via CFTR and Na^+-dependent fluid absorption via the epithelial Na^+ channel ENaC. ENaC is the limiting step for sodium absorption in the lungs (Rossier *et al.*, 2002) as recognized from mice with a targeted disruption of the α-subunit of ENaC that die immediately after birth because of defective fluid clearance in the lungs (Hummler *et al.*, 1996). Increased activity of ENaC in CF patients partially depends on the missing negative regulation of ENaC activity by wild-type CFTR (Reddy *et al.*, 1999; Stutts *et al.*, 1995). Furthermore, near-silent epithelial Na^+ channels can be activated by endogenous serine proteases and proteases released from neutrophils (Caldwell *et al.*, 2004, 2005). Fluid protease inhibitors at the airway surface also indirectly influence ENaC activity, so that alterations in the serine-protease/ protease-inhibitor balance contribute to excessive Na^+ absorption in cystic fibrosis (Myerburg *et al.*, 2006). The concept that airway surface dehydration can produce CF-like lung disease was confirmed in mice that overexpress the βENaC subunit. These mice die of CF-like symptoms including mucus obstruction with neutrophilic infiltration due to impaired Cl^- secretion and increased Na^+ absorption (Mall *et al.*, 2004). This concept is further supported by the observation of gain-of-function mutations in the gene encoding the β-subunit of ENaC (*SCNN1B*) in some patients with CF-like symptoms (sweat Cl^- greater than 60 mmol/l and CF-like pulmonary infections) without mutations in the coding regions of *CFTR* (Sheridan *et al.*, 2005).

Surprisingly, the lung phenotype is not reproduced in mouse models deficient for CFTR which rather show a severe intestinal phenotype (Clarke et al., 1992). In the colon, CFTR localizes to the apical side of crypt epithelial cells, consistent with a role for chloride secretion. The $[Cl^-]_i$ of colonic crypt cells is above equilibrium, so that opening of CFTR leads to the efflux of chloride and the passive secretion of water, its loss resulting in impaired chloride and water secretion with thick feces. Conversely, inadequately strong activation of CFTR by the drastic increase in cAMP that is elicited by cholera toxin leads to severe diarrhea, which can be alleviated in animal models by blocking the basolateral K^+ conductance (Rufo et al., 1996).

It was speculated that other chloride channels as, for example, the ubiquitously expressed voltage-gated chloride channel ClC-2 may partially compensate for CFTR. However, mice with a targeted disruption of both CFTR and ClC-2 showed neither lung nor pancreatic disease and survived even better compared with CFTR knockout mice (Zdebik et al., 2004). ClC-2 has been reported to be present in apical (Mohammad-Panah et al., 2001; Murray et al., 1995) or basolateral (Catalan et al., 2004; Murray et al., 1995) membranes of various epithelia, with much better support for the basolateral localization. In colonic epithelia, cells at the luminal surface express a Cl^-/HCO_3^- exchanger and the Na^+/H^+ exchanger NHE3 at the apical membrane for NaCl reabsorption. In surface epithelia of mouse colon, ClC-2 is expressed at the basolateral membrane (Catalan et al., 2004; Pena-Munzenmayer et al., 2005; Zdebik et al., 2004); it can be concluded that chloride probably passes the basolateral membrane through ClC-2 in these cells. In epithelial cells at the crypt base, chloride is taken up basolaterally by the cation chloride cotransporter NKCC1 and secreted apically through CFTR channels, KCNQ1/KCNE3 heteromeric K^+ channels serving as a basolateral K^+ recycling pathway (Schroeder et al., 2000).

In the pancreatic gland, CFTR activity allows chloride efflux and thereby enhances HCO_3^- secretion via apical anion exchange (Fig. 5.1). CFTR itself was also shown to conduct HCO_3^- and may exert an activating effect on HCO_3^- exchangers (Choi et al., 2001; Ko et al., 2004).

Therapy mainly aims at the treatment and prevention of pulmonary complications using antibiotics, bronchodilators, anti-inflammatory agents, mucolytic agents, and chest physiotherapy. In selected patients, lung or heart/lung transplantation is performed, although according to a recent study no prolongation of life by means of lung transplantation should be expected (Liou et al., 2007). New insights into the pathophysiology of CF have provided the means to develop new rational therapeutic strategies. One approach is the inhibition of ENaC to increase the airway surface fluid by inhibiting Na^+ reabsorption. However, the effects of aerosol treatment with amiloride, an inhibitor of ENaC, and more potent compounds like benzamil and phenamil in vivo have been disappointing. This may be explained by a limited bioavailability (Hirsh *et al.*, 2004), but more suitable derivates of amiloride are being developed (Hirsh *et al.*, 2008). Another rational is the inhibition of endogenous serine proteases that are expressed in the airways and promote Na^+ reabsorption via a proteolytic activation of ENaC (Donaldson *et al.*, 2002). Neutrophil elastase is another serine protease and present in CF airways at remarkable concentrations due to airway infection and inflammation (Konstan *et al.*, 1994) and contributes to excessive airway Na^+ absorption and thereby exacerbates the airway pathology. A novel neutrophil elastase inhibitor prevented activation of ENaC and may hence be a promising candidate for clinical trails (Harris *et al.*, 2007).

    Efficient delivery and sustained expression of intact CFTR in an already compromised epithelium are a major problem in gene replacement strategies. Both nonreplicating viral vectors and DNA–lipid complexes have been tried but did not prove clinically useful. Easier may be the suppression of premature stop mutations, for example, by amino glycoside antibiotics. However, this strategy will be applicable only in patients that have at least one allele with a premature stop codon (approximately 10%). Although gentamicin application to the nasal mucosa of suitable patients caused a significant increase in full-length CFTR as detected by immunocytochemistry (Wilschanski *et al.*, 2003), there was no improvement in a recent metacenter study (Clancy *et al.*, 2007). Because of the complex multidomain structure of the CFTR protein, the proper folding and the intracellular trafficking of CFTR are very critical even for the wild-type protein. The $\Delta F508$ variant, although being synthesized, fails completely to be transported to the cell surface and results in a severe pulmonary phenotype in the homozygous state (Cheng *et al.*, 1990). Hence, high-throughput screens to identify compounds that promote the processing of this CFTR variant in a cellular *in vitro* system are underway. Several compounds have been identified and may be candidates for clinical trias (Pedemonte *et al.*, 2005).

B. Channelopathies affecting renal function

The maintenance of electrolyte homeostasis and water balance is vital for the organism. The kidneys are the main players to preserve the right "milieu interne." For this purpose, kidney glomeruli filter approximately 120 ml of plasma per minute, whereas blood cells and plasma proteins are retained in the capillaries of the filtration apparatus. Low molecular weight proteins, salts, and water pass into the more than 1.2 million renal tubules. Many components of this primary filtrate are reabsorbed as the fluid passes the tubular epithelium, a process that does not only depend on the structural integrity of the kidney, but also depend on a complex interplay between ion channels, pumps, and transporters.

The tubular apparatus can be subdivided into parts with different functions (Fig. 5.2) that are also reflected by expression of different sets of ion channels and ion transporters.

1. Focal segmental glomerulosclerosis

The filtration apparatus consists of a bundle of capillaries surrounded by the glomerular basement membrane and a layer of podocytes, the interdigitating foot processes of podocytes covering the basement membrane. The filtration slits between podocyte processes are closed by a thin diaphragmatic structure with pores ranging in size between 5 and 15 nm. The filtration of positively charged solutes is favored owing to the negatively charged glycoproteins in the basement membrane and in podocytes, filtration slits, and slit diaphragms.

If the glomerular filter is destroyed, a nephrotic syndrome is the consequence in many cases. This disorder is defined by the clinical trias of edema, proteinuria, and hypercholesterolemia. It is often caused by focal segmental glomerulosclerosis (FSGS) as the underlying disorder, both in children (up to 20%) and in adults (up to 35% of cases) (Haas *et al.*, 1995). Primary (idiopathic), secondary, or familial forms can be distinguished. Secondary FSGS can occur in diseases like

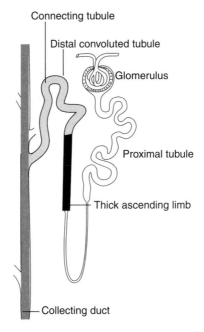

Figure 5.2. Schematic drawing of the nephron.

human immunodeficiency virus infection, heroin abuse, and sickle cell disease. The familial forms of FSGS are transmitted as autosomal-dominant and autosomal-recessive traits. The causative mutations affect genes that are thought to be involved in podocyte signaling or the structural filtration barrier of the glomerulus (for review, see Reidy and Kaskel, 2007). One of the genes identified to date is TRPC6 (Winn et al., 2005) that encodes a member of the TRP cation channel family. The TRP family comprises heterotetrameric nonselective cation channels (Nilius et al., 2005) that mediate diverse biological functions such as ion homeostasis, mechanosensation, cell growth, and vasoregulation. They are permeable to monovalent cations as well as calcium with a relative lack of selectivity. The growing spectrum of diseases caused by mutations in TRP channels also includes hypomagnesemia with secondary hypocalcemia (TRPM6) (Schlingmann et al., 2002), mucolipidosis type IV (TRPML1) (Bargal et al., 2000), and polycystic kidney disease (TRPP2) (Peters et al., 1993), disorders that will be discussed later in this chapter.

TRPC6 locates to the podocyte cell membrane where it partially colocalizes with other podocyte proteins such as nephrin and podocin. The initially described TRPC6 mutation P112Q lies within the first ankyrin-binding domain at the amino-terminal part of the protein and is associated with a particularly aggressive clinical course. Although this TRPC6 mutation was shown to promote calcium influx into the cytoplasm of podocytes, it is still unclear how this relates to FSGS. It is possible that increased intracellular calcium modifies the contractile structure of podocyte foot processes, thereby affecting ultrafiltration. As podocytes of nephrin-deficient mice overexpress mislocalized TRPC6 proteins (Reiser et al., 2005), it was suggested that TRPC6 may be a component of a signaling complex that is located at the slit diaphragm. In this model, abnormal TRPC6 expression may cause structural changes in the slit diaphragm and thereby lead to proteinuria and glomerulosclerosis. Indeed, cultured podocytes overexpressing TRPC6 lost actin stress fibers (Moller et al., 2007). As TRPC6 knockout mice are hypertensive (Dietrich et al., 2005), it was suggested that TRPC6 may be involved in the control of vascular smooth muscle tone. In lung tissues and pulmonary arterial smooth-muscle cells from patients with idiopathic pulmonary arterial hypertension, mRNA and protein expression of TRPC6 (and TRPC3) were increased compared with those from normotensive probands or patients with secondary pulmonary hypertension (Weissmann et al., 2006). Inhibition of TRPC6 expression with TRPC6 small interfering RNA attenuated proliferation of pulmonary arterial smooth-muscle cells of patients with pulmonary hypertension (Yu et al., 2004).

Current treatment of FSGS is symptomatic by blocking the renin–angiotensin system with angiotensin-converting enzyme (ACE) inhibitors or angiotensin-receptor blockers. In the future, a more specific strategy may aim at blocking TRPC6. However, this is a difficult challenge as TRPC6 is highly homologous to TRPC3 and TRPC7 and expressed in a wide variety of human tissues.

2. Autosomal-dominant polycystic kidney disease

In autosomal-dominant polycystic kidney disease (ADPKD), the structure of the kidney is destroyed by cysts that arise from renal tubules. This process ultimately leads to renal failure in half of the affected patients. It is one of the most prevalent, potentially lethal, monogenic disorders with a prevalence of approximately one in 1000. A substantial variability of severity of renal disease and other extrarenal manifestations has been observed even within the same family. The diagnosis of ADPKD is established, if individuals at risk for the disease younger than 30 years of age show two or more unilateral or bilateral cysts. Extrarenal manifestations include cysts in other organs such as the liver and pancreas (up to 75%), vascular abnormalities such as intracranial aneurysms (approximately 10%) and dilatation of the aortic root, and mitral valve prolapse in up to 25% of affected individuals.

In the majority of patients with ADPKD, mutations in either *PKD1* (Reeders *et al.*, 1985) encoding the membrane protein polycystin-1 or *PKD2* (Peters *et al.*, 1993) encoding polycystin-2/TRPP2 can be identified. Sequence analysis of *PKD1* and *PKD2* results in the detection of a disease-causing mutation in approximately 85% of patients with *PKD1* being by far the more frequent cause. No clear correlation between genotype and phenotype has been found, but overall the clinical presentation is more severe in PKD1. Most patients with *PKD1* mutations develop renal failure at the age of 70 years, whereas in about 50% of individuals with *PKD2* mutations adequate renal function is found at that age. Despite of a high penetrance, the clinical inter- and intrafamilial variability is considerable and includes antenatal manifestations with an incidence of less than 1%. The variable clinical manifestations in siblings compared with monozygotic twins may be explained by the existence of genetic modifier genes (Persu *et al.*, 2004). Although a two-hit mechanism with a germline and a somatic inactivation of the two alleles has been proposed (Watnick *et al.*, 1998), which could explain the focal formation of cysts, haploinsufficiency is thought to account for vascular manifestations (Qian *et al.*, 2003).

The protein products of *PKD1* and *PKD2*, polycystin-1 and polycystin-2/TRPP2, are membrane proteins that probably form a functional complex (Low *et al.*, 2006). The polycystin-1/TRPP2 complex localizes to the primary cilium of tubular epithelial cells. This single hair-like organelle projects into the lumen of the tubule. It was postulated that polycystin-1 is essential for the translocation of polycystin-2/TRPP2 into the plasma membrane, where both proteins should coassemble to form a cation-selective ion channel (Hanaoka *et al.*, 2000; Qian *et al.*, 1997). However, polycystin-2/TRPP2 belongs to the TRP channel superfamily and in itself may form a calcium-activated, high-conductance ER channel that is permeable to calcium (Koulen *et al.*, 2002). Polycystin-2/TRPP2 also interacts physically with TRPC1 (Bai *et al.*, 2008), a store-operated Ca^{2+}

channel that has recently been shown to be stretch activated. Although mice deficient for polycystin-1 had morphologically normal cilia, no flow-induced calcium response of renal epithelial cells was detectable (Nauli et al., 2003). Mice deficient for polycystin-2/TRPP2 show laterality defects in addition to polycystic kidneys (Pennekamp et al., 2002), supporting the notion that polycystin-2/TRPP2 also plays a role for left/right asymmetry. It emerges that the polycystin complex may serve as a mechanosensor that translates luminal flow into calcium influx mediated by polycystin-2/TRPP2. The ensuing increase of intracellular cAMP levels results in activation of various intracellular pathways that can trigger cell proliferation.

A better understanding of the cellular events may lead to more specific treatment strategies in ADPKD in the future. Promising results were obtained with vasopressin receptor 2 antagonists that were selected because of the effects of vasopressin on cellular cAMP levels. One compound (OPC31260) substantially reduced renal cAMP concentration and inhibited cyst progression in animal models of ADPKD (Gattone et al., 2003).

3. Tubular disorders

The proximal tubule of the kidney reabsorbs ~60% of filtered NaCl and fluid and most of the filtered amino acids and glucose. Another ~25% of filtered NaCl are reabsorbed in the thick ascending limb of Henle's loop (TAL). As the TAL epithelial cells are virtually impermeable for water, this generates a hypertonic kidney medulla. During antidiuresis, this osmotic gradient is used to reabsorb water in the collecting duct. For this purpose, aquaporin-2 water channels are inserted into the plasma membrane of collecting duct principal cells in the presence of antidiuretic hormone to allow water reabsorption. For the fine tuning of renal acid secretion, α-intercalated cells in distal convoluted tubules, connecting tubules, and collecting ducts are equipped with a luminal H^+-ATPase. Sodium reabsorption is controlled by principal cells that express the luminal epithelial sodium channel ENaC under the control of aldosterone.

a. Bartter syndrome

Bartter syndrome is a group of closely related hereditary tubulopathies, which are characterized by renal salt-wasting, hypokalemic metabolic alkalosis, and hyperreninemic hyperaldosteronism with normal blood pressure. Several clinical variants were described, all forms following an autosomal-recessive trait: a severe antenatal form with and without deafness, often resulting in premature delivery due to polyhydramnios and a failure to thrive, and the classic Bartter syndromes, which occur in infancy or early childhood. Patients suffering from Gitelman syndrome are also characterized by renal salt loss and hypotension, but have a less severe phenotype compared with Bartter syndrome patients.

Gitelman syndrome is caused by a defective apical NaCl reabsorption in the distal convoluted tubule caused by loss-of-function mutations in the gene encoding the NaCl-cotransporter NCCT (*SLC4A3*) (Simon *et al.*, 1996c), whereas renal salt loss in Bartter syndrome results from impaired transepithelial transport in the thick ascending limb of the loop of Henle. The molecules involved have been identified and mutations in either component of this transport system can cause Bartter syndrome (Fig. 5.3).

The apical $Na^+–K^+–2Cl^-$-cotransporter (NKCC2/SLC12A1) is essential for NaCl uptake from the lumen into cells of the thick ascending limb of Henle's loop. It is mutated in severe antenatal forms without deafness (Simon *et al.*, 1996a). As NKCC2 transports K^+ into the cell, K^+ must be recycled apically via the K^+ channel ROMK/Kir1.1 (encoded by *KCNJ1*). Hence, its loss of function also results in Bartter syndrome (Simon *et al.*, 1996b). Na^+ leaves the cell actively via the basolateral Na^+/K^+-ATPase, whereas Cl^- diffuses through basolateral ClC-Kb chloride channels (*CLCNKB*) following its electrochemical gradient (Simon *et al.*, 1997). Like in the inner ear, ClC-Kb needs the β-subunit Barttin (Estévez *et al.*, 2001) for proper targeting. This β-subunit is encoded by

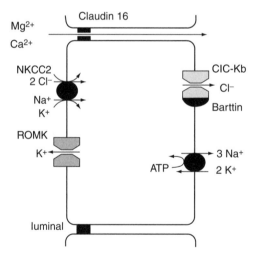

Figure 5.3. Salt uptake in the thick ascending limb. In cells of the thick ascending limb of Henle's loop, apical NKCC2-cotransporters drive Cl^- uptake. The K^+ channel ROMK is indispensable for apical K^+ recycling. Cl^- exits through basolateral channels formed by ClC-Kb and its β-subunit barttin. Mutations in all four corresponding genes can cause Bartter syndrome. Driven by a luminal-positive voltage, Mg^{2+} is reabsorbed in the thick ascending limb through a paracellular pathway that involves claudin 16. In contrast, in the distal convoluted tubule, Mg^{2+} is transported transcellularly by TRPM6 (see Fig. 6.4). Mutations in either gene can result in renal Mg^{2+} loss.

the *BSND* gene and is structurally unrelated to the CLC family. It was identified as the disease-causing gene in a large pedigree with the antenatal variant of Bartter syndrome with deafness (Bartter type IV) (Birkenhäger *et al.*, 2001).

A coding *CLCNKB* polymorphism (ClC-KbT481S) drastically increases ClC-Kb Cl^- channel activity (Jeck *et al.*, 2004a) and occurs with a prevalence of 20% in Caucasians and 40% in Africans. Expression of the variant channel may decrease the cytosolic Cl^- concentration and thus enhance the driving force and transport rate of $Na^+–2Cl^-–K^+$-cotransport. As a result, the gene variant might lead to enhanced salt resorption in the kidney. The gene variant has been associated with increased blood pressure in one study (Jeck *et al.*, 2004b), but this could not be reproduced in later studies (Fava *et al.*, 2007; Kokubo *et al.*, 2005; Speirs *et al.*, 2005).

Recently, the impact of rare alleles for *SLC12A1* (NKCC1), *SLC12A3* (NCC), and *KCNJ1* (ROMK) on blood pressure has been addressed in the Framingham Heart Study offspring cohort that has been followed up for cardiovascular risk factors for more than 30 years. Rare heterozygous mutations have been identified that are associated with clinically significant blood pressure reduction (Ji *et al.*, 2008).

Patients with antenatal Bartter syndrome have excessive levels of prostaglandin E_2, which is thought to promote salt and water loss. Indeed, indomethacin that blocks the formation of prostaglandins by inhibition of both cyclooxygenase-1 and -2 alleviates the symptoms in patients with antenatal Bartter syndrome. More recent data suggest that inhibition of cyclooxygenase-2 may be sufficient for therapy (Reinalter *et al.*, 2002).

b. Disorders of renal NaCl balance related to ENaC

The apical sodium entry step into principal cells of the distal tubule is mediated by the sodium-selective, amiloride-sensitive ion channel ENaC that also has an important role in the pathophysiology of CF as already discussed earlier. It is driven by the electrochemical potential difference across the apical membrane from the tubular lumen into the cell. The exit step into the interstitium is catalyzed by the basolateral Na^+, K^+-ATPase (Fig. 5.4). Na^+ absorption is always accompanied by a corresponding osmotic uptake of water. Thus, Na^+ absorption leads to an expansion of blood volume and raises the blood pressure. Indeed, mutations of the epithelial Na^+ channel ENaC can lead either to hereditary hypotension if sodium reabsorption is impaired or to hypertension if sodium reabsorption is increased. The antidiuretic hormone (ADH) enhances Na^+ and H_2O uptake by insertion of ENaC and aquaporin-2 water channels into the plasma membrane.

ENaC comprises three homologous subunits (α, β, γ) with 30% homology at the protein level. Each subunit has two transmembrane domains with short cytoplasmic amino- and carboxy-termini and a large extracellular loop. ENaC subunits share a number of conserved domains important for channel function.

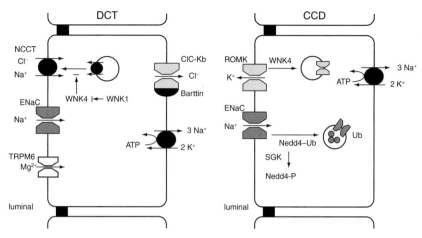

Figure 5.4. Salt uptake in the distal convoluted tubule (DCT, *left*) and the cortical collecting duct (CCD, *right*). Salt uptake in the DCT is mediated by NCCT. An increase of NaCl uptake through NCC contributes to pseudohypoaldosteronism (PHA) type 2, which is caused by mutations of WNK1 or WNK4. Both kinases differentially affect the apical insertion of NCCT NaCl-cotransporters. In the cortical collecting duct, WNK4 stimulates the endocytosis of ROMK. PHA-2-specific WNK4 mutations further increase this effect. The predicted decrease of ROMK in the apical membrane leads to decreased K^+ secretion. Na^+ absorption in the cortical collecting duct is mediated by the epithelial sodium channel ENaC. Its loss of function causes PHA type 1, whereas gain-of-function mutations underlie hereditary hypertension.

ENaC is almost certainly a heterotrimer, as indicated by the crystal structure of the related ASIC1 ion channel (Jasti *et al.*, 2007). Amiloride, a very effective blocker of ENaC, is thought to act as an open channel blocker, which plugs the pore and physically prevents ion flux (Schild *et al.*, 1997).

Pseudohypoaldosteronism type 1 (PHA-1) is an inherited disease characterized by severe neonatal salt-wasting, hyperkalemia, metabolic acidosis, and its unresponsiveness to mineralocorticoid hormones. An autosomal-dominant form is caused by a loss-of-function mutation in the mineralocorticoid receptor (Geller *et al.*, 1998), whereas the more severe recessive form is caused by loss-of-function mutations in α-, β-, or γ- ENaC subunits (Chang *et al.*, 1996). It is associated with marked hypotension and dehydration of newborns and infants due to excessive loss of Na^+. Missense mutations causing PHA-1 are found in critically important domains.

In contrast, a rare form of autosomal-dominant salt-sensitive hypertension with secondary hypokalemia and metabolic acidosis (Liddle syndrome) is caused by gain of function of ENaC-dependent Na^+ reabsorption. Mutations at a specific position of the ENaC β- or γ-subunit have been identified (Hansson *et al.*, 1995a,b) that is part of a motif characterized by the typical amino acid sequence (PPxY). Mutations in this "PY-motif" lead to an increased amount of the protein in the plasma membrane by preventing the interaction of ENaC with WW-domain containing ubiquitin ligases like Nedd-4. Ubiquitination leads to an enhanced rate of endocytosis and degradation of the channel complex (Staub et al., 1996, 1997). The lack of this repressor activity leads to an increased cell surface expression of ENaC Na^+ channels, entailing a constitutively enhanced Na^+ reabsorption that results in severe hypertension. These mutations therefore cause a gain of function, explaining the dominant inheritance pattern of Liddle syndrome.

Each subunit of ENaC has been inactivated in mice. Although embryonic and fetal development was not impaired, knockout animals for each subunit died within the first days after birth. αENaC knockout pups died from impaired fluid clearance in the lungs (Hummler et al., 1996), showing that ENaC plays a critical role in the adaptation of the newborn lung to air breathing. βENaC (McDonald et al., 1999) and γENaC (Barker et al., 1998) knockout pups have low urinary potassium and high urinary sodium concentrations and most likely died of hyperkalemia within the first days after birth. As γENaC-deficient mice cleared their lung fluid more slowly than control littermates, γENaC also seems to be involved in neonatal lung fluid clearance.

c. Nephrogenic diabetes insipidus

Many epithelial cells display high water permeability, although the ability of water to pass hydrophobic lipid bilayers is very limited. Hence, the existence of water transport systems in many cell membranes was postulated. The 28-kDa integral membrane protein of red cells and renal tubules was the first member of a large protein family identified as water channels and is now known as aquaporin channel-forming integral protein (CHIP) or aquaporin-1 (AQP1) (Preston and Agre, 1991; Smith and Agre, 1991).

From the cDNA sequence, the existence of two tandem repeats with three bilayer-spanning α-helices was deduced (Preston et al., 1992). Based on site directed mutagenesis an "hourglass model" with the six bilayer-spanning α-helices surrounding the aqueous pore was proposed (Jung et al., 1994). The protein oligomerizes in the cell membrane to build a tetramer with four individual water channel pores (Jung et al., 1994). AQP1 is abundantly expressed in the kidney, where it localizes to the proximal tubule, the thin descending limb of Henle's loop and the vasa recta in the outer medulla. As an AQP1 epitope is associated with a blood group antigen, some human individuals that lack AQP1 have been identified by chance. These individuals are clinically not affected, but a significant defect in their maximal urine concentration capacity could be demonstrated during fluid deprivation (King et al., 2001). In vitro proximal tubule microperfusion and in vivo micropuncture in mice with a targeted disruption of AQP1 revealed drastically diminished water permeability in the proximal tubule, the thin descending limb of Henle, and the outer medullary descending vasa recta (Schnermann et al., 1998). This resulted in defective near-isosmolar fluid absorption in the proximal tubule and defective countercurrent multiplication (Schnermann et al., 1998).

Loss of AQP2 function results in a much more pronounced phenotype (Deen et al., 1994) with excessive renal water loss patients voiding up to 20 l of urine per day. This nephrogenic diabetes insipidus is caused by the inability of the renal collecting ducts to absorb water. In healthy individuals, AQP2 is exclusively expressed in principal cells of the connecting tubule and collecting duct. Its insertion into the plasma membrane is regulated by the antidiuretic hormone/vasopressin thus allowing water reabsorption form the lumen during antidiuresis. The inheritance is usually autosomal recessive but maybe autosomal dominant in some patients, as a dominant-negative effect leading to the missorting of the protein was shown in a knockin mouse model (Sohara et al., 2006). However, much more frequently nephrogenic diabetes insipidus follows an X-linked recessive trait and is then caused by mutation of the gene encoding the vasopressin V2 receptor (Rosenthal et al., 1992).

The major intrinsic protein (MIP) of lens fiber is another member of the aquaporin gene family. Originally, it was cloned from bovine lens cells in 1984 (Gorin et al., 1984) and was thought to serve a role as a junctional protein (Gorin et al., 1984). It took several years until MIP was recognized as a member of the aquaporin gene family and is since referred to as AQP0. Identification of mutations in mice (Shiels and Bassnett, 1996) and humans (Berry et al., 2000) with severe, dominantly inherited cataracts points to an important role of the MIP/AQP0 protein for the lens. MIP/AQP0 exhibits rather low water permeability and was speculated to serve a dual role as a water channel and a structural protein (Fotiadis et al., 2000).

d. Familial hypomagnesemia

Magnesium is essential for many cellular functions like bone stabilization, neurotransmission, muscular relaxation, cardiac action potential, and enzymatic reactions. It freely passes the glomerulus and is not effectively reabsorbed in the proximal tubule. A paracellular uptake pathway exists in the thick ascending limb of Henle's loop, whereas reabsorption is realized transcellularly in the distal tubule.

The molecular correlate of this transcellular pathway was identified by linkage analysis in three large inbred kindreds from Israel with familial hypomagnesaemia. Subsequently, mutations were identified in TRPM6 (Schlingmann et al., 2002; Walder et al., 2002), a member of the TRP cation channel family. TRPM6 localizes to the luminal site of cells of the distal convoluted tubule and allows the entry of magnesium along its electrochemical gradient. The subsequent transport across the basolateral membrane remains elusive at the molecular level. Mutations in epidermal growth factor can also cause familial hypomagnesemia (Groenestege et al., 2007). This may be explained by regulation of TRPM6 activity by epidermal growth factor. Cetuximab, a drug used in treatment of cancer, acts as an inhibitor of the epidermal growth factor receptor and entails hypomagnesemia as a known side effect, which may be due to the inhibition of epidermal growth factor signaling (Groenestege *et al.*, 2007).

Interestingly, a mutation in the γ-subunit of the Na^+, K^+-ATPase segregated with hypomagnesemia in a dominant pedigree (Meij *et al.*, 2000). This putatively dominant-negative mutation is predicted to affect the trans-epithelial voltage difference and thereby paracellular Mg^{2+} transport.

The paracellular uptake pathway in the thick ascending limb also depends on claudin 16 [also known as paracellin-1 (PCLN-1)] that belongs to the claudin family of tight junction proteins. This transmembrane protein was identified by a positional cloning approach in a large consanguineous family suffering of renal Mg^{2+} wasting with autosomal-recessively inherited mutations in paracellin-1 (Simon *et al.*, 1999).

e. X-linked hypercalciuric nephrolithiasis (Dent's disease)

Dent's disease refers to a group of X-linked hereditary renal tubular disorders characterized by hypercalciuric nephrolithiasis caused by mutations in the *CLCN5* gene (Lloyd *et al.*, 1996). Although these disorders are allelic and are all characterized by progressive proximal renal tubulopathy with hypercalciuria, low molecular weight proteinuria, and nephrocalcinosis, they vary in degree of severity and were originally reported as separate disorders. Some female carriers may have asymptomatic proteinuria, hypercalciuria, or hypophosphatemia only. In the meantime many different *CLCN5* mutations in Dent's disease patients have been reported, including missense mutations and truncations. In addition to low molecular weight proteinuria with hypercalciuria, hyperphosphaturia, which leads to kidney stones in the majority of patients, there is a variable presence of other symptoms of proximal tubular dysfunction such as glucosuria and aminoaciduria (Wrong *et al.*, 1994).

CLCN5 codes for the voltage-dependent Cl^-/H^+ exchanger ClC-5 (Picollo and Pusch, 2005; Scheel *et al.*, 2005). The kidney is the major expression site of ClC-5 (Steinmeyer *et al.*, 1995). It is most prominently expressed in the proximal tubule and in intercalated cells of the collecting duct. While the function of ClC-5 in intercalated cells is not yet clear, its role in the proximal tubule has been elucidated by the use of knockout mouse models (Piwon *et al.*, 2000; Wang *et al.*, 2000). ClC-5 is expressed in endosomes of proximal tubules (Fig. 5.5), where it colocalizes with the H^+-ATPase (Günther *et al.*, 1998). The knockout of ClC-5 led to low molecular weight proteinuria and hyperphosphaturia. The proteinuria is a consequence of an impaired endocytosis in proximal tubule epithelial cells, which extends to fluid-phase endocytosis, receptor-mediated endocytosis, and the endocytosis of integral plasma membrane proteins such as the Na^+–phosphate cotransporter NaPi-IIa (Piwon *et al.*, 2000). However, endocytosis is not completely abolished. The amount of

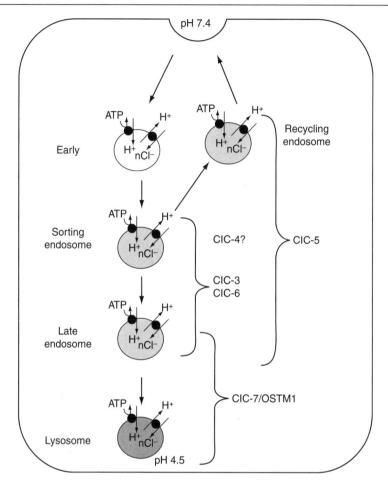

Figure 5.5. Model for the acidification of the endosomal/lysosomal pathway. Vesicles of the endo-
somal/lysosomal pathway are acidified by V-type H⁺-ATPases. The voltage over the
vesicular membrane that would be generated by this process is neutralized by chloride
entering the vesicles through channels/transporters of the CLC type; for example, ClC-5
in endosomes and ClC-7 in lysosomes. The localization to specific compartments,
however, is less clear for ClC-4.

the endocytotic receptor megalin, which mediates the uptake of a wide variety of
proteins and other substrates, is reduced in the absence of ClC-5 (Piwon et al.,
2000), suggesting that ClC-5 may have a role in recycling this receptor back to
the apical plasma membrane. Isolated renal cortical endosomes had a lower rate
and extent of acidification compared with wild-type (Günther et al., 2003). Thus
ClC-5 may provide an electrical shunt for the vesicular proton pump that is
necessary for the efficient acidification of endosomes. The resulting defect of

endocytosis also extends to the reabsorption of filtered parathormone (PTH) that normally binds to megalin. An increased luminal PTH concentration in more distal parts of the proximal tubule is the consequence. This will stimulate luminal PTH receptors and in turn the endocytosis of NaPi-IIa, thereby explaining hyperphosphaturia (Piwon et al., 2000). Together with changes in proximal tubular handling of vitamin D that are also secondary to the primary defect of endocytosis, these findings provide an explanation for the pathogenesis of kidney stones in Dent's disease. ClC-5 has a putative PY-motif in its carboxy-terminal tail, which is also found in other channels expressed along the tubule such as ENaC. This may suggest a role of ubiquitin ligases in the regulation of protein turnover by ubiquitination and endocytosis (Hryciw et al., 2004; Schwake et al., 2001).

It should be mentioned that certain mutations in OCRL, which encodes an inositol-polyphosphate phosphatase, can lead to symptoms indistinguishable from Dent's disease caused by CLCN5 mutations (Hoopes et al., 2005), although most OCRL mutations cause Lowe's syndrome that includes ocular and CNS symptoms (Attree et al., 1992).

C. Hearing loss and channelopathies

Hearing loss already affects many newborns, almost 1/1000 children being born with severe hearing impairment. In the general population, the prevalence of hearing loss increases with age. Depending on the site of the underlying pathology it is classified as either conductive (middle ear), sensorineural (inner ear), or a combination of both. It is quantified in decibels (dB) above the threshold that normal individuals need to perceive a tone of a given frequency 50% of the time. Hearing is considered normal if an individual's thresholds are within 15 dB of control probands.

Auditory deprivation during early childhood results in poor speech production and communication skills. However, cognitive impairments in persons with hereditary hearing loss are not intrinsically linked to the cause of deafness. Hence, early auditory intervention with hearing aid devices or cochlear implantation is important. Assessment of cognitive skills in individuals with connexin 26-related deafness revealed a normal IQ and normal reading performance after early cochlear implantation (Taitelbaum-Swead et al., 2006).

In more than half of the cases with prelingual deafness (acquired before onset of speech), a genetic origin can be assumed that is most often inherited as an autosomal-recessive trait. The disorder DFNB1, caused by mutations in the GJB2 gene (which encodes the gap junction protein connexin 26) and the GJB6 gene (encoding connexin 30) accounts for 50% of autosomal-recessive nonsyndromic hearing loss. The carrier rate in the general population for a recessive deafness-causing GJB2 mutation is about 1 in 30. Many more less common monogenic forms of hearing loss have been identified and have resulted in a considerable insight into the physiology and pathology of hearing. Not surprisingly, many genes are related to ion homeostasis. In the following, we will not limit this review to channelopathies of the inner ear that are affecting epithelial transport, but also other ion channel diseases for the sake of giving a more complete picture.

The inner ear is the sensory system for sound, motion, and gravity. It consists of the cochlea, the vestibular labyrinth, and the endolymphatic sac. The cochlea transduces sound into electrical signals that are conveyed to the brain. It contains three fluid compartments enclosed by epithelial cells: the scala tympani, the scala media, and the scala tympani. The apical membranes of sensory hair cells of the organ of Corti are in contact with the endolymph of the scala media. The endolymph has a very high K^+, but low Na^+ concentration. Moreover, the

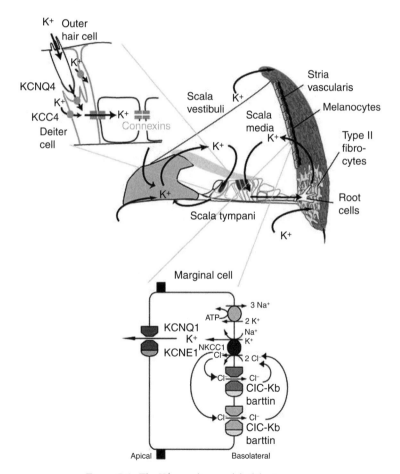

Figure 5.6. The K^+ recycling model of the inner ear.

potential of this compartment is held at $+90$ to $+100$ mV with respect to the remaining extracellular space. This unusual ion composition and potential is established by a specialized epithelium at the lateral wall of the scala media, the stria vascularis. This two-layered epithelium is composed of a marginal cell layer and a layer of basal cells that is penetrated by a capillary network. At the apical membrane of marginal cells, K^+ is secreted into the endolymph through KCNQ1/ KCNE1 K^+ channels (Fig. 5.6). This secretion is possible because the intracellular potential of marginal cells is even more positive (by a few mV) than the endolymph. Homozygous loss of either KCNQ1 (Kv7.1) or its β-subunit KCNE1 (also known as minK or IsK) underlie the congenital deafness of the Jervell–Lange–Nielsen syndrome (Neyroud et al., 1997; Schulze-Bahr et al., 1997; Tyson et al., 1997). This syndrome also includes potentially life-threatening cardiac arrhythmia as the channel complex is also important for the repolarization of the cardiac action potential. Mice deficient for Kcne1 have a collapse of Reissner's membrane that separates the scala media from the scala vestibuli, indicating that fluid secretion into the endolymph depends on K^+ secretion by the stria vascularis (Vetter et al., 1996). The K^+ ions that are secreted at the apical membrane of marginal cells is accumulated over the basolateral membrane by the combined activity of the Na^+, K^+-ATPase and the Na–K–2Cl-cotransporter NKCC1. As expected by this model, NKCC1-deficient mice show a collapse of Reissner's membrane (Delpire et al., 1999) similar to that observed with a KO of the apical KCNQ1/KCNE1 K^+ channel. Cl^- ions taken up by the NKCC1-cotransporter must be recycled across the basolateral membrane. This occurs through ClC-Ka/Barttin and ClC-Kb/Barttin chloride channels and explains why mutations in BSND, the gene encoding Barttin, result in deafness: The loss of the common β-subunit impairs transport through both ClC-Ka/barttin and ClC-Kb/barttin chloride channels (Birkenhäger et al., 2001; Estévez et al., 2001). In contrast, mutations in ClC-Kb resulting in renal salt loss (which is also observed with a loss of Barttin) can be compensated by ClC-Ka/Barttin. Supporting this hypothesis, in rare patients mutations in both ClC-Ka and ClC-Kb result in deafness (Nozu et al., 2008; Schlingmann et al., 2004). Barttin is structurally unrelated to the CLC family and was identified by a positional cloning strategy in a large family with deafness associated with a salt-loosing disorder of the kidney (Birkenhäger et al., 2001), as discussed previously.

The high potassium concentration and the positive voltage of the endolymph are needed to provide the appropriate driving force for depolarizing K^+ entry into sensory hair cells upon an acoustic stimulus. Apical mechanosensitive ion channels of so far unknown molecular identity are expressed in the stereocilia of sensory hair cells that open upon deflection. The depolarization of hair cells triggers the opening of voltage-gated Ca^{2+} channels. The ensuing rise in intracellular Ca^{2+} leads to the exocytosis of synaptic vesicles in sensory inner hair cells (IHCs), which depolarize postsynaptic membranes of neurons of the spiral ganglion. The axons of these neurons convey the acoustic information to the brain via the auditory nerve. The sound-evoked changes in membrane potential of sensory outer hair cells (OHCs), by contrast, primarily serve to drive directly mechanical responses of those contractile cells. The motor protein prestin, which is highly expressed in the lateral membranes of OHCs, changes the length of these cells and amplifies the mechanical vibrations in the organ of Corti. Potassium ions that have entered sensory hair cells through apical mechanosensitive channels must leave these cells through their basal membrane. This involves the voltage-gated K^+ channel KCNQ4 (Kv7.4), which is primarily expressed in OHCs (Kubisch et al., 1999), and the large-conductance Ca^{2+}-activated K^+ channel KCNMA1 ($K_{Ca}1.1$), a BK channel (Kros et al., 1998). Heterozygous mutations of KCNQ4 can result in autosomal-dominant progressive hearing loss (DFNA2) (Kubisch et al., 1999). Most human KCNQ4 mutations exert a dominant-negative effect on coexpressed wild-type KCNQ4 subunits in a heterologous expression system. Mice lacking KCNQ4 or expressing a dominant-negative mutation identified in human patients initially display only slight hearing impairment that is correlated with a depolarization of OHCs and, to a lesser extent, IHCs (Kharkovets et al., 2006). Reduced otoacoustic emissions revealed a moderate impairment of OHC function before the onset of degeneration. The selective degeneration of OHCs, but not IHCs, explains the slowly progressing hearing loss. It was slower in mice heterozygous for the dominant-negative mutant than in the complete KO, indicating that the predicted 6% of normal channel function in those heterozygotes can significantly delay OHC degeneration (Kharkovets et al., 2006). Similarly, mice lacking KCNMA1 that is expressed in both IHCs and OHCs develop normal hearing but then show progressive hearing loss (Rüttiger et al., 2004), indicating that KCNMA1 like KCNQ4 is nonessential for basic IHC function. Similar to KCNQ4-deficient mice, mice lacking KCNMA1 lose OHCs, but not IHCs, during the progressive loss of hearing. Besides, gain-of-function mutations in KCNMA1 were identified in human epilepsy and paroxysmal movement disorder (Du et al., 2005). In the future, these animal models may be useful to test pharmacological strategies (e.g., K^+ channel openers) to delay the progression of deafness.

    Hair cells maintain their negative resting membrane potential via basal K^+ channels in conjunction with high cytosolic and low extracellular K^+ concentrations. Uncontrolled increases of the K^+ concentration in the extracellular fluid bathing the basal portion of these cells are expected to depolarize the membrane potential and the responsiveness of hair cells. Hence, K^+ must be efficiently removed after leaving OHCs, to a large extent by uptake into supporting Deiters' cells. This uptake is probably mediated by the KCl-cotransporters KCC4 (SC12A7) and KCC3 (SC12A6), both of which are expressed in Deiters' cells. Disruption of KCC4 in the mouse leads to early onset degeneration of hair cells (Boettger et al., 2002), whereas hearing impairment progresses slowly in the absence of KCC3 (Boettger *et al.*, 2003). By coupling K^+ to Cl^- flux, KCl-cotransporters generally operate close to the combined electrochemical equilibrium. Although KCl-cotransport is best known for mediating KCl efflux, it may also allow influx with slight changes of extra- and intracellular ion concentrations. The potential and potassium concentrations of the scala media and endolymph allows K^+ influx into hair cells through apical mechanosensitive channels, its efflux through basal KCNQ and KCNMA1 channels, as well as its uptake into supporting cells passively without any input of metabolic energy in those cells. The "powerhouse" driving this system is the stria vascularis that provides the favorable electrochemical gradients. Therefore, the stria has to be highly vascularized, whereas the organ of Corti can be free of blood vessels, a feature that seems essentially for its function in mechanoelectrical transduction.

The supporting Deiters' cells are connected to root cells in the spiral ligament by an epithelial gap junction system, which provides a cytoplasmic route for K^+ diffusion. After exiting from root cells to the extracellular space, K^+ is taken up by type II fibrocytes. These are coupled to type I and type III fibrocytes in a fibrocyte gap junction system that also includes the basal and intermediate cells of the stria vascularis. K^+ leaving intermediate cells is then taken up by marginal cells and is secreted through apical K^+ channels. After having left root cells, K^+ is accumulated by type II fibrocytes through NKCC1 and the Na^+, K^+-ATPase (Kikuchi *et al.*, 2000). Type II cells express the K^+-accumulating NKCC1, but not KCC3, and the reverse is true for type I and III fibrocytes. This may create a K^+ gradient within the fibrocytes gap junction system. The exit into the space between intermediate and marginal cells of the stria occurs through Kir4.1 K^+ channels (Ando and Takeuchi, 1999). This electrogenic exit generates the endocochlear potential. If KCC3 were expressed in type II fibrocytes or strial intermediate cells, its transport activity would interfere with K^+ recycling and the generation of the endocochlear potential, respectively. Type II fibrocytes (lacking KCC3) were often preserved in KO mice that already displayed a severe loss of type I and III fibrocytes (which express KCC3) (Boettger *et al.*, 2003).

At least three connexin genes, *GJB2*, *GJB3*, and *GJB6*, which encode components of the inner ear gap junction system, are mutated in human genetic deafness (Denoyelle *et al.*, 1998; Grifa *et al.*, 1999; Kelsell *et al.*, 1997). In humans, mutations of connexin 26 (*GJB2*) and 30 (*GJB6*), as already mentioned above, are by far the most frequent etiologies of nonsyndromic genetic deafness. Mice that lack GJB6 were deaf and failed to develop an endocochlear potential but had normal endolymphatic K^+ concentrations (Teubner *et al.*, 2003). It is assumed that the limited coupling mediated by the remaining GJB2 (and other connexins) is sufficient for K^+ cycling, but results in leakiness of strial capillaries and breakdown of the endocochlear potential. For several deafness-causing GJB2 and GJB6 variants the transfer of organic molecules is impaired, but not the ionic coupling. This implies that these mutations do not affect K^+ recycling but rather metabolic coupling and glutamate buffering. It is intriguing that the loss of function of either GJB2 or GJB6 leads to deafness rather than simply being compensated by the remaining homomeric gap junctions. The finding that hearing of mice lacking GJB6 can be rescued by overexpression of GJB2 (Ahmad et al., 2007) suggests that the biophysical differences between hetero- and homomeric GJB2 gap junctions are less important than the fact that loss of GJB6 leads to a loss of GJB2 expression and a reduction in intercellular coupling. Whereas connexin 26 knockout mice are embryonic lethal, mice with a selective disruption of Gjb2 in cells of the epithelial compartment of the inner ear are viable and show no developmental defect of the inner ear (Cohen-Salmon et al., 2002). However, starting on postnatal day 14 cell death of supporting cells of IHCs was observed that subsequently extended to the cochlear epithelial network and sensory hair cells (Cohen-Salmon et al., 2002).

D. Mucolipidosis type IV

Mucolipidosis type IV is an autosomal-recessive neurodegenerative lysosomal storage disorder characterized by psychomotor retardation and ophthalmologic abnormalities that has been reported for the first time in 1974 (Berman et al., 1974). Other clinical features are achlorhydria and hypergastrinemia. It has been classified as a lysosomal disorder because of accumulation of laminated, membranous materials that can be identified in lysosomes of various tissues from patients (Merin et al., 1975). As some other lysosomal storage disorders (Gaucher, Tay-Sachs, Niemann-Pick disease types A and B), it is quite common in the Ashkenazi population with a frequency of around one in 40,000 and a heterozygote frequency of 1/100. Most patients are diagnosed during early infancy because of severe mental retardation and ophthalmologic manifestations such as opacity of the cornea (Bach, 2001). Despite of early manifestation little deterioration usually occurs during the first decades of life.

The underlying genetic defect was not identified until recently (Bargal et al., 2000). The gene, MCOLN1 codes for TRP-ML1 (or mucolipin), a member of the TRP-ML subfamily of TRP channels (Bassi et al., 2000; Sun et al., 2000). TRP-ML1 was reported to be a Ca^{2+} channel (LaPlante et al., 2002), or an outwardly rectifying monovalent cation channel regulated by either Ca^{2+} (Cantiello et al., 2005) or pH (Raychowdhury et al., 2004). Pathogenic mutations alter cellular localization or ion selectivity and permeability of TRP-ML1 (Kiselyov et al., 2005). As TRP-ML1 is probably a lysosomal ion channel, it is expected to regulate lysosomal ion content. However, there are still many open questions in the pathogenesis of the disease. From earlier biochemical studies a defect in the endocytosis process of membranous components was suggested

(Bargal and Bach, 1997). Recent work reports that TRP-ML1 may limit lysosomal acidification by providing a lysosomal H^+ leak pathway (Soyombo *et al.*, 2006). The acidification of the lysosomal lumen is mediated by a vacuolar H^+ pump (Beyenbach and Wieczorek, 2006) and probably members of the CLC family of Cl^- transporters (Jentsch *et al.*, 2005). A murine model for mucolipidosis type IV accurately replicates the phenotype of patients (Venugopal *et al.*, 2007). In support of an over acidification of lysosomes in cells deficient of Trp-Mll, the phenotype could be reversed by dissipating lysosomal pH (Soyombo *et al.*, 2006). TRP-ML1 has also been suggested to modulate the formation of lysosomes by mediating fusion of lysosomes with late endosomes or fission of lysosomes from hybrid organelles (Treusch *et al.*, 2004). In *Caenorhabditis elegans*, a loss-of-function mutation in the mucolipin-1 homolog, cup5, results in an enhanced rate of uptake of fluid-phase markers, decreased degradation of endocytosed protein, and accumulation of large vacuoles (Fares and Greenwald, 2001).

E. Osteopetrosis and lysosomal storage disease

In osteopetrosis an imbalance between the formation and breakdown of bone results in inadequately dense bones. Several types of osteopetrosis of varying severity are distinguished with diverse side effects. The severe autosomal-recessive variant becomes symptomatic in early infancy. Absence of the bone marrow cavity results in severe anemia and thrombocytopenia. Sclerosis of the skull base can result in blindness, facial palsy and hearing loss due to cranial nerve compression. Additional facultative features are dental abnormalities, odontomas, mandibular osteomyelitis, and hypocalcaemia with tetanic seizures and secondary hyperparathyroidism. Without treatment maximal lifespan is around 10 years. The prevalence of the milder variant Albers–Schönberg disease has been estimated to be approximately 1:100,000 and follows an autosomal-dominant trait. Autosomal-recessive osteopetrosis is probably even less common. Depending on the population studied, penetrance ranges from 60% to 90% (Waguespack *et al.*, 2003).

Mice with a targeted disruption of the *Clcn7* gene, which codes for an intracellular member of the family of CLC chloride channels, develop severe osteopetrosis, retinal degeneration and neurodegeneration (Kornak *et al.*, 2001). Screening of 12 patients with infantile osteopetrosis for mutations in the human *CLCN7* gene resulted in the identification of compound heterozygosity for a nonsense (Q555X) and a missense (R762Q) mutation in the *CLCN7* gene in one of the patients (Kornak *et al.*, 2001). In the meantime, the spectrum of *CLCN7*-related osteopetrosis also includes Albers–Schönberg disease (Cleiren *et al.*, 2001). Mutations in Albers–Schönberg disease are present in a heterozygous state and presumably exert a dominant-negative effect on the coexpressed product of the normal allele. Indeed, so far only missense, but not nonsense mutations of *CLCN7* have been identified in the dominant form of the disease, probably yielding proteins that can still associate with the WT protein encoded by the normal allele in heterozygous patients. Because of the dimeric structure of CLC channels and transporters, about 25% of transport function should be left upon a 1:1 coexpression of dominant negative and WT alleles. Hence, osteopetrosis in Albers–Schönberg disease is less severe and needs several years to decades to develop. It is usually not associated with blindness or lysosomal storage disease.

The localization of ClC-7 to the acid-secreting ruffled border of osteoclasts, the cells involved in bone degradation, gave a first hint how dysfunction of ClC-7 may affect bone resorption (Kornak *et al.*, 2001). The ruffled border is formed by the exocytotic insertion of H^+-ATPase-containing vesicles of late endosomal/lysosomal origin. ClC-7 is coinserted with the H^+-ATPase into this membrane and may serve, like ClC-5 in endosomes, as a shunt for the acidification of the resorption lacuna (Fig. 5.7). Acidification of the resorption lacuna is crucial for the chemical dissolution of inorganic bone material, as well as for the activity of secreted lysosomal enzymes that degrade the organic bone matrix. In a cell culture system, ClC-7 KO osteoclasts still attached to ivory but were unable to acidify the resorption lacuna and to degrade the bone surrogate (Kornak *et al.*, 2001). Supporting this model, mutations in the H^+-ATPase also can result in osteopetrosis (Frattini *et al.*, 2000; Kornak *et al.*, 2000).

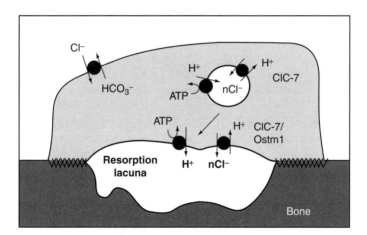

Figure 5.7. Model for the acidification of the resorption lacuna. In bone-resorbing osteoclasts, proton pumps and ClC-7/Ostm1 Cl^- transporters are inserted into the "ruffled border" membrane facing the bone surface. Thereby, the resorption lacuna obtains its characteristic acidic pH, which is essential for the dissolution of the mineral phase as well as the enzymatic degradation of the organic bone matrix.

Importantly, osteopetrosis caused by a total loss of ClC-7 is complicated by neurodegeneration, but not if it is caused by mutations in the H^+-ATPase. (Kasper et al., 2005). This points to a more general role of ClC-7 in the acidification of intracellular organelles. Indeed, mice deficient for ClC-7 also develop severe neurodegeneration (Kasper et al., 2005). Neurodegeneration was also observed in mice deficient for ClC-3 (Stobrawa et al., 2001) and ClC-6 (Poët et al., 2006), two more intracellular CLC chloride transporters. However, the severity and the morphological and biochemical characteristics of the degeneration differed significantly between the genotypes. In mice lacking ClC-6 or ClC-7, neurons displayed intracellular, electron-dense deposits that stained for lysosomal marker proteins and the subunit c of ATP synthase, a protein typically accumulated in a subset of human lysosomal storage diseases called neuronal ceroid lipofuscinosis (NCL). Storage occurred throughout neuronal cell bodies in ClC-7 KO mice (Kasper et al., 2005), whereas it accumulated specifically in initial axon segments of mice lacking ClC-6 (Poët et al., 2006). In comparison, no severe intraneuronal storage was observed in ClC-3 KO mice (Kasper et al., 2005; Stobrawa et al., 2001), although an independent mouse model deficient of ClC-3 was reported to display NCL-like features (Yoshikawa et al., 2002). The neuronal cell loss was severe in mice lacking ClC-3, leading to a complete loss of the hippocampus in adult mice (Stobrawa et al., 2001). Mice deficient for ClC-7 also displayed severe neuronal cell loss and died after about 2 months, even if their osteopetrosis was cured by the transgenic expression of ClC-7 in osteoclasts (Kasper et al., 2005), whereas mice deficient for ClC-3 survived for more than a year in a mixed genetic background. Neuronal cell loss was nearly absent in ClC-6 KO mice, which showed nearly normal lifespan and only mild neurological abnormalities that included a reduced sensitivity to pain (Poët et al., 2006). Although the combination of severe osteopetrosis with lysosomal storage disease may also occur in human patients homozygous for severe *CLCN7* mutations, it is currently unclear whether human mutations in *CLCN3* or *CLCN6* may cause phenotypes similar to ClC-3 and ClC-6 KO mice.

The lysosomal pathologies observed with a disruption of either ClC-6 or ClC-7 raise the question of the subcellular localization of these CLC proteins. Immunocytochemistry, immunogold labeling, and subcellular fractionation demonstrated the presence of ClC-7 in lysosomes and late endosomes (Kasper et al., 2005; Kornak et al., 2001; Lange et al., 2006). Staining for ClC-7 and the late endosomal–lysosomal protein lamp-1 overlapped almost completely, whereas there was only a partial overlap with ClC-6 (Poët et al., 2006). These studies demonstrated that ClC-7 is the only CLC protein that is significantly expressed on lysosomes. ClC-6 and ClC-3 may both be expressed predominantly on late endosomes (Poët et al., 2006), a conclusion also supported by localizing epitope-tagged CLCs in transfected cells (Suzuki et al., 2006). A prelysosomal localization of ClC-3 and ClC-6 is further suggested by the partial shift of both proteins, but not of ClC-4, into lysosomal fractions in mice deficient for ClC-7 (Poët et al., 2006). In neuronal cells, ClC-3 is additionally expressed on synaptic vesicles (Stobrawa et al., 2001).

The lysosomal storage disease observed with the disruption of the lysosomal ClC-7 transporter suggested that lysosomal acidification might be impaired, similar to the lack of acidification of the resorption lacuna (Kornak et al., 2001). Surprisingly, however, careful ratiometric measurements of lysosomal pH of ClC-7 KO neurons and fibroblasts did not show any difference of lysosomal pH as compared with WT (Kasper et al., 2005). Less surprisingly (because of the late endosomal localization of ClC-6), the pH of $Clcn6^{-/-}$ lysosomes was unchanged (Poët et al., 2006). It was therefore speculated that a change in lysosomal Cl^- concentration, and not pH, may cause the observed lysosomal storage (Jentsch, 2007).

Recently, the functional link between CLCN7 and another gene associated with severe autosomal-recessive osteopetrosis, OSTM1, was resolved (Lange et al., 2006). A homozygous mutation in OSTM1 was identified by linkage analysis and sequence analysis of genes in the respective chromosomal position in a spontaneous mouse mutant, grey lethal, that develops severe osteopetrosis and has a characteristic grey fur color in an agouti genetic background. Screening of severely affected human patients with severe autosomal-recessive osteopetrosis detected a homozygous mutation in one out of 19 patients with infantile malignant osteopetrosis (Chalhoub et al., 2003). Only a few more patients have been identified to date (Pangrazio et al., 2006; Ramirez et al., 2004). OSTM1 encodes a type I membrane protein (Ostm1) with a heavily glycosylated amino-terminal portion and a short cytoplasmic tail (Lange et al., 2006). Ostm1 colocalizes with ClC-7 in lysosomes and in the ruffled border of osteoclasts. ClC-7 could be coimmunoprecipitated with Ostm1 and vice versa, identifying Ostm1 as a novel β-subunit of ClC-7 (Lange et al., 2006). Ostm1 needs ClC-7 to be targeted to lysosomes, whereas ClC-7 reaches lysosomes also in the absence of Ostm1. Importantly, the stability of either protein depends on the coexpression with its partner. This interaction provides an explanation for the pathogenesis of this rare severe variant of osteopetrosis (Lange et al., 2006): the pathology observed upon a loss of Ostm1 may be entirely due to the ensuing instability of the ClC-7 chloride transporter. Indeed, grey lethal mice do not only resemble ClC-7 knockout mice in their bone phenotype, but also display a similar lysosomal storage disease. Like in ClC-7 KO mice, lysosomal pH of cells derived from grey lethal mice was not changed (Lange et al., 2006). Although the grey hair of either mouse model is not yet fully understood, it may be related to a dysfunction of melanosomes, which are lysosome-related organelles.

Bone sclerosis and secondary symptoms such as bone marrow failure, extramedullary hematopoiesis, and impairments caused by nerve compression can be reversed or at least prevented by hematopoietic stem cell transplantation. Primary neurological and retinal manifestation, however, are independent of the bone disease and therefore cannot be cured by this approach.

III. CONCLUDING REMARKS

Mutations in ion channels and transporters are now known to underlie many different human pathologies, which are not at all restricted to disturbances of excitable cells. These human diseases and knockout mouse models have shed considerable light on the physiological importance of ion transport. Modern, massively parallel sequencing of patient and normal cohorts will undoubtedly uncover more mutations in ion transport proteins, which may in some cases also be associated with beneficial effects, as recently shown by a study investigating three renal ion transporters in a large cohort (Ji et al., 2008). Many more surprising insights into the diverse roles of ion transport can be expected.

References

Ahmad, S., Tang, W., Chang, Q., Qu, Y., Hibshman, J., Li, Y., Sohl, G., Willecke, K., Chen, P., and Lin, X. (2007). Restoration of connexin26 protein level in the cochlea completely rescues hearing in a mouse model of human connexin30-linked deafness. Proc. Natl Acad. Sci. USA 104, 1337–1341.

Ando, M., and Takeuchi, S. (1999). Immunological identification of an inward rectifier K^+ channel (Kir4.1) in the intermediate cell (melanocyte) of the cochlear stria vascularis of gerbils and rats. Cell Tissue Res. 298, 179–183.

Attree, O., Olivos, I. M., Okabe, I., Bailey, L. C., Nelson, D. L., Lewis, R. A., McInnes, R. R., and Nussbaum, R. L. (1992). The Lowe's oculocerebrorenal syndrome gene encodes a protein highly homologous to inositol polyphosphate-5-phosphatase. Nature 358, 239–242.

Bach, G. (2001). Mucolipidosis type IV. Mol. Genet. Metab. 73, 197–203.

Bai, C. X., Giamarchi, A., Rodat-Despoix, L., Padilla, F., Downs, T., Tsiokas, L., and Delmas, P. (2008). Formation of a new receptor-operated channel by heteromeric assembly of TRPP2 and TRPC1 subunits. EMBO Rep. 9, 472–479.

Bargal, R., and Bach, G. (1997). Mucolipidosis type IV: Abnormal transport of lipids to lysosomes. J. Inherit. Metab. Dis. 20, 625–632.

Bargal, R., Avidan, N., Ben-Asher, E., Olender, Z., Zeigler, M., Frumkin, A., Raas-Rothschild, A., Glusman, G., Lancet, D., and Bach, G. (2000). Identification of the gene causing mucolipidosis type IV. Nat. Genet. 26, 118–123.

Barker, P. M., Nguyen, M. S., Gatzy, J. T., Grubb, B., Norman, H., Hummler, E., Rossier, B., Boucher, R. C., and Koller, B. (1998). Role of gammaENaC subunit in lung liquid clearance and electrolyte balance in newborn mice. Insights into perinatal adaptation and pseudohypoaldosteronism. J. Clin. Invest. 102, 1634–1640.

Bassi, M. T., Manzoni, M., Monti, E., Pizzo, M. T., Ballabio, A., and Borsani, G. (2000). Cloning of the gene encoding a novel integral membrane protein, mucolipidin-and identification of the two major founder mutations causing mucolipidosis type IV. Am. J. Hum. Genet. 67, 1110–1120.

Berman, E. R., Livni, N., Shapira, E., Merin, S., and Levij, I. S. (1974). Congenital corneal clouding with abnormal systemic storage bodies: A new variant of mucolipidosis. *J. Pediatr.* **84,** 519–526.

Berry, V., Francis, P., Kaushal, S., Moore, A., and Bhattacharya, S. (2000). Missense mutations in MIP underlie autosomal dominant 'polymorphic' and lamellar cataracts linked to 12q. *Nat. Genet.* **25,** 15–17.

Beyenbach, K. W., and Wieczorek, H. (2006). The V-type H$^+$ ATPase: Molecular structure and function, physiological roles and regulation. *J. Exp. Biol.* **209,** 577–589.

Birkenhäger, R., Otto, E., Schurmann, M. J., Vollmer, M., Ruf, E. M., Maier-Lutz, I., Beekmann, F., Fekete, A., Omran, H., Feldmann, D., *et al.* (2001). Mutation of BSND causes Bartter syndrome with sensorineural deafness and kidney failure. *Nat. Genet.* **29,** 310–314.

Boettger, T., Hübner, C. A., Maier, H., Rust, M. B., Beck, F. X., and Jentsch, T. J. (2002). Deafness and renal tubular acidosis in mice lacking the K–Cl co-transporter Kcc4. *Nature* **416,** 874–878.

Boettger, T., Rust, M. B., Maier, H., Seidenbecher, T., Schweizer, M., Keating, D. J., Faulhaber, J., Ehmke, H., Pfeffer, C., Scheel, O., *et al.* (2003). Loss of K–Cl co-transporter KCC3 causes deafness, neurodegeneration and reduced seizure threshold. *EMBO J.* **22,** 5422–5434.

Caldwell, R. A., Boucher, R. C., and Stutts, M. J. (2004). Serine protease activation of near-silent epithelial Na$^+$ channels. *Am. J. Physiol.* **286,** C190–C194.

Caldwell, R. A., Boucher, R. C., and Stutts, M. J. (2005). Neutrophil elastase activates near-silent epithelial Na$^+$ channels and increases airway epithelial Na$^+$ transport. *Am. J. Physiol. Lung Cell. Mol. Physiol.* **288,** L813–L819.

Cantiello, H. F., Montalbetti, N., Goldmann, W. H., Raychowdhury, M. K., Gonzalez-Perrett, S., Timpanaro, G. A., and Chasan, B. (2005). Cation channel activity of mucolipin-1: The effect of calcium. *Pflugers Arch.* **451,** 304–312.

Catalan, M., Niemeyer, M. I., Cid, L. P., and Sepulveda, F. V. (2004). Basolateral ClC-2 chloride channels in surface colon epithelium: Regulation by a direct effect of intracellular chloride. *Gastroenterology* **126,** 1104–1114.

Chalhoub, N., Benachenhou, N., Rajapurohitam, V., Pata, M., Ferron, M., Frattini, A., Villa, A., and Vacher, J. (2003). Grey-lethal mutation induces severe malignant autosomal recessive osteopetrosis in mouse and human. *Nat. Med.* **9,** 399–406.

Chang, S. S., Gründer, S., Hanukoglu, A., Rosler, A., Mathew, P. M., Hanukoglu, I., Schild, L., Lu, Y., Shimkets, R. A., Nelson-Williams, C., *et al.* (1996). Mutations in subunits of the epithelial sodium channel cause salt wasting with hyperkalaemic acidosis, pseudohypoaldosteronism type 1. *Nat. Genet.* **12,** 248–253.

Cheng, S. H., Gregory, R. J., Marshall, J., Paul, S., Souza, D. W., White, G. A., O'Riordan, C. R., and Smith, A. E. (1990). Defective intracellular transport and processing of CFTR is the molecular basis of most cystic fibrosis. *Cell* **63,** 827–834.

Choi, J. Y., Muallem, D., Kiselyov, K., Lee, M. G., Thomas, P. J., and Muallem, S. (2001). Aberrant CFTR-dependent HCO$_3^-$ transport in mutations associated with cystic fibrosis. *Nature* **410,** 94–97.

Clancy, J. P., Rowe, S. M., Bebok, Z., Aitken, M. L., Gibson, R., Zeitlin, P., Berclaz, P., Moss, R., Knowles, M. R., Oster, R. A., *et al.* (2007). No detectable improvements in cystic fibrosis transmembrane conductance regulator by nasal aminoglycosides in patients with cystic fibrosis with stop mutations. *Am. J. Respir. Cell Mol. Biol.* **37,** 57–66.

Clarke, L. L., Grubb, B. R., Gabriel, S. E., Smithies, O., Koller, B. H., and Boucher, R. C. (1992). Defective epithelial chloride transport in a gene-targeted mouse model of cystic fibrosis. *Science* **257,** 1125–1128.

Cleiren, E., Benichou, O., Van Hul, E., Gram, J., Bollerslev, J., Singer, F. R., Beaverson, K., Aledo, A., Whyte, M. P., Yoneyama, T., *et al.* (2001). Albers–Schonberg disease (autosomal dominant osteopetrosis, type II) results from mutations in the ClCN7 chloride channel gene. *Hum. Mol. Genet.* **10,** 2861–2867.

Cohen-Salmon, M., Ott, T., Michel, V., Hardelin, J. P., Perfettini, I., Eybalin, M., Wu, T., Marcus, D. C., Wangemann, P., Willecke, K., and Petit, C. (2002). Targeted ablation of connexin26 in the inner ear epithelial gap junction network causes hearing impairment and cell death. *Curr. Biol.* **12**, 1106–1111.

Deen, P. M., Verdijk, M. A., Knoers, N. V., Wieringa, B., Monnens, L. A., van Os, C. H., and van Oost, B. A. (1994). Requirement of human renal water channel aquaporin-2 for vasopressin-dependent concentration of urine. *Science* **264**, 92–95.

Delpire, E., Lu, J., England, R., Dull, C., and Thorne, T. (1999). Deafness and imbalance associated with inactivation of the secretory Na–K–2Cl co-transporter. *Nat. Genet.* **22**, 192–195.

Denoyelle, F., Lina-Granade, G., Plauchu, H., Bruzzone, R., Chaib, H., Levi-Acobas, F., Weil, D., and Petit, C. (1998). Connexin 26 gene linked to a dominant deafness. *Nature* **393**, 319–320.

Dietrich, A., Mederos, Y. S. M., Gollasch, M., Gross, V., Storch, U., Dubrovska, G., Obst, M., Yildirim, E., Salanova, B., Kalwa, H., *et al.* (2005). Increased vascular smooth muscle contractility in TRPC6$^{-/-}$ mice. *Mol. Cell. Biol.* **25**, 6980–6989.

Donaldson, S. H., Hirsh, A., Li, D. C., Holloway, G., Chao, J., Boucher, R. C., and Gabriel, S. E. (2002). Regulation of the epithelial sodium channel by serine proteases in human airways. *J. Biol. Chem.* **277**, 8338–8345.

Du, W., Bautista, J. F., Yang, H., Diez-Sampedro, A., You, S. A., Wang, L., Kotagal, P., Luders, H. O., Shi, J., Cui, J., *et al.* (2005). Calcium-sensitive potassium channelopathy in human epilepsy and paroxysmal movement disorder. *Nat. Genet.* **37**, 733–738.

Estévez, R., Boettger, T., Stein, V., Birkenhäger, R., Otto, E., Hildebrandt, F., and Jentsch, T. J. (2001). Barttin is a Cl$^-$ channel beta-subunit crucial for renal Cl$^-$ reabsorption and inner ear K$^+$ secretion. *Nature* **414**, 558–561.

Fares, H., and Greenwald, I. (2001). Genetic analysis of endocytosis in *Caenorhabditis elegans*: Coelomocyte uptake defective mutants. *Genetics* **159**, 133–145.

Fava, C., Montagnana, M., Almgren, P., Rosberg, L., Guidi, G. C., Berglund, G., and Melander, O. (2007). The functional variant of the CLC-Kb channel T481S is not associated with blood pressure or hypertension in Swedes. *J. Hypertens.* **25**, 111–116.

Fotiadis, D., Hasler, L., Muller, D. J., Stahlberg, H., Kistler, J., and Engel, A. (2000). Surface tongue-and-groove contours on lens MIP facilitate cell-to-cell adherence. *J. Mol. Biol.* **300**, 779–789.

Frattini, A., Orchard, P. J., Sobacchi, C., Giliani, S., Abinun, M., Mattsson, J. P., Keeling, D. J., Andersson, A. K., Wallbrandt, P., Zecca, L., *et al.* (2000). Defects in TCIRG1 subunit of the vacuolar proton pump are responsible for a subset of human autosomal recessive osteopetrosis. *Nat. Genet.* **25**, 343–346.

Gadsby, D. C., Vergani, P., and Csanady, L. (2006). The ABC protein turned chloride channel whose failure causes cystic fibrosis. *Nature* **440**, 477–483.

Gattone, V. H., II, Wang, X., Harris, P. C., and Torres, V. E. (2003). Inhibition of renal cystic disease development and progression by a vasopressin V2 receptor antagonist. *Nat. Med.* **9**, 1323–1326.

Geller, D. S., Rodriguez-Soriano, J., Vallo Boado, A., Schifter, S., Bayer, M., Chang, S. S., and Lifton, R. P. (1998). Mutations in the mineralocorticoid receptor gene cause autosomal dominant pseudohypoaldosteronism type I. *Nat. Genet.* **19**, 279–281.

Gorin, M. B., Yancey, S. B., Cline, J., Revel, J. P., and Horwitz, J. (1984). The major intrinsic protein (MIP) of the bovine lens fiber membrane: Characterization and structure based on cDNA cloning. *Cell* **39**, 49–59.

Grifa, A., Wagner, C. A., D'Ambrosio, L., Melchionda, S., Bernardi, F., Lopez-Bigas, N., Rabionet, R., Arbones, M., Monica, M. D., Estivill, X., *et al.* (1999). Mutations in GJB6 cause nonsyndromic autosomal dominant deafness at DFNA3 locus. *Nat. Genet.* **23**, 16–18.

Groenestege, W. M., Thebault, S., van der Wijst, J., van den Berg, D., Janssen, R., Tejpar, S., van den Heuvel, L. P., van Cutsem, E., Hoenderop, J. G., Knoers, N. V., and Bindels, R. J. (2007). Impaired basolateral sorting of pro-EGF causes isolated recessive renal hypomagnesemia. *J. Clin. Invest.* **117**, 2260–2267.

Guggino, W. B. (1999). Cystic fibrosis and the salt controversy. *Cell* **96,** 607–610.

Günther, W., Luchow, A., Cluzeaud, F., Vandewalle, A., and Jentsch, T. J. (1998). ClC-5, the chloride channel mutated in Dent's disease, colocalizes with the proton pump in endocytotically active kidney cells. *Proc. Natl Acad. Sci. USA* **95,** 8075–8080.

Günther, W., Piwon, N., and Jentsch, T. J. (2003). The ClC-5 chloride channel knock-out mouse— An animal model for Dent's disease. *Pflugers Arch.* **445,** 456–462.

Haas, M., Spargo, B. H., and Coventry, S. (1995). Increasing incidence of focal-segmental glomerulosclerosis among adult nephropathies: A 20-year renal biopsy study. *Am. J. Kidney Dis.* **26,** 740–750.

Hanaoka, K., Qian, F., Boletta, A., Bhunia, A. K., Piontek, K., Tsiokas, L., Sukhatme, V. P., Guggino, W. B., and Germino, G. G. (2000). Co-assembly of polycystin-1 and -2 produces unique cation-permeable currents. *Nature* **408,** 990–994.

Hansson, J. H., Nelson-Williams, C., Suzuki, H., Schild, L., Shimkets, R., Lu, Y., Canessa, C., Iwasaki, T., Rossier, B., and Lifton, R. P. (1995a). Hypertension caused by a truncated epithelial sodium channel gamma subunit: Genetic heterogeneity of Liddle syndrome. *Nat. Genet.* **11,** 76–82.

Hansson, J. H., Schild, L., Lu, Y., Wilson, T. A., Gautschi, I., Shimkets, R., Nelson-Williams, C., Rossier, B. C., and Lifton, R. P. (1995b). A de novo missense mutation of the beta subunit of the epithelial sodium channel causes hypertension and Liddle syndrome, identifying a proline-rich segment critical for regulation of channel activity. *Proc. Natl Acad. Sci. USA* **92,** 11495–11499.

Harris, M., Firsov, D., Vuagniaux, G., Stutts, M. J., and Rossier, B. C. (2007). A novel neutrophil elastase inhibitor prevents elastase activation and surface cleavage of the epithelial sodium channel expressed in *Xenopus laevis* oocytes. *J. Biol. Chem.* **282,** 58–64.

Hirsh, A. J., Sabater, J. R., Zamurs, A., Smith, R. T., Paradiso, A. M., Hopkins, S., Abraham, W. M., and Boucher, R. C. (2004). Evaluation of second generation amiloride analogs as therapy for cystic fibrosis lung disease. *J. Pharmacol. Exp. Ther.* **311,** 929–938.

Hirsh, A. J., Zhang, J., Zamurs, A., Fleegle, J., Thelin, W. R., Caldwell, R. A., Sabater, J. R., Abraham, W. M., Donowitz, M., Cha, B., *et al.* (2008). Pharmacological properties of N-(3,5-diamino-6-chloropyrazine-2-carbonyl)-N'-4-[4-(2,3-dihydroxypropoxy)phenyl]butyl-guanidine methanesulfonate (552-02), a novel epithelial sodium channel blocker with potential clinical efficacy for cystic fibrosis lung disease. *J. Pharmacol. Exp. Ther.* **325,** 77–88.

Hoopes, R. R., Jr, Shrimpton, A. E., Knohl, S. J., Hueber, P., Hoppe, B., Matyus, J., Simckes, A., Tasic, V., Toenshoff, B., Suchy, S. F., *et al.* (2005). Dent disease with mutations in OCRL1. *Am. J. Hum. Genet.* **76,** 260–267.

Hryciw, D. H., Ekberg, J., Lee, A., Lensink, I. L., Kumar, S., Guggino, W. B., Cook, D. I., Pollock, C. A., and Poronnik, P. (2004). Nedd4-2 functionally interacts with ClC-5: Involvement in constitutive albumin endocytosis in proximal tubule cells. *J. Biol. Chem.* **279,** 54996–55007.

Hummler, E., Barker, P., Gatzy, J., Beermann, F., Verdumo, C., Schmidt, A., Boucher, R., and Rossier, B. C. (1996). Early death due to defective neonatal lung liquid clearance in alpha-ENaC-deficient mice. *Nat. Genet.* **12,** 325–328.

Jasti, J., Furukawa, H., Gonzales, E. B., and Gouaux, E. (2007). Structure of acid-sensing ion channel 1 at 1.9 A resolution and low pH. *Nature* **449,** 316–323.

Jeck, N., Waldegger, P., Doroszewicz, J., Seyberth, H., and Waldegger, S. (2004a). A common sequence variation of the CLCNKB gene strongly activates ClC-Kb chloride channel activity. *Kidney Int.* **65,** 190–197.

Jeck, N., Waldegger, S., Lampert, A., Boehmer, C., Waldegger, P., Lang, P. A., Wissinger, B., Friedrich, B., Risler, T., Moehle, R., *et al.* (2004b). Activating mutation of the renal epithelial chloride channel ClC-Kb predisposing to hypertension. *Hypertension* **43,** 1175–1181.

Jentsch, T. J. (2007). Chloride and the endosomal–lysosomal pathway: Emerging roles of CLC chloride transporters. *J. Physiol.* **578,** 633–640.

Jentsch, T. J., Poët, M., Fuhrmann, J. C., and Zdebik, A. A. (2005). Physiological functions of CLC Cl⁻ channels gleaned from human genetic disease and mouse models. *Annu. Rev. Physiol.* **67**, 779–807.

Ji, W., Foo, J. N., O'Roak, B. J., Zhao, H., Larson, M. G., Simon, D. B., Newton-Cheh, C., State, M. W., Levy, D., and Lifton, R. P. (2008). Rare independent mutations in renal salt handling genes contribute to blood pressure variation. *Nat. Genet.* **40**, 592–599.

Jung, J. S., Preston, G. M., Smith, B. L., Guggino, W. B., and Agre, P. (1994). Molecular structure of the water channel through aquaporin CHIP. The hourglass model. *J. Biol. Chem.* **269**, 14648–14654.

Kasper, D., Planells-Cases, R., Fuhrmann, J. C., Scheel, O., Zeitz, O., Ruether, K., Schmitt, A., Poët, M., Steinfeld, R., Schweizer, M., *et al.* (2005). Loss of the chloride channel ClC-7 leads to lysosomal storage disease and neurodegeneration. *EMBO J.* **24**, 1079–1091.

Kelsell, D. P., Dunlop, J., Stevens, H. P., Lench, N. J., Liang, J. N., Parry, G., Mueller, R. F., and Leigh, I. M. (1997). Connexin 26 mutations in hereditary non-syndromic sensorineural deafness. *Nature* **387**, 80–83.

Kharkovets, T., Dedek, K., Maier, H., Schweizer, M., Khimich, D., Nouvian, R., Vardanyan, V., Leuwer, R., Moser, T., and Jentsch, T. J. (2006). Mice with altered KCNQ4 K⁺ channels implicate sensory outer hair cells in human progressive deafness. *EMBO J.* **25**, 642–652.

Kikuchi, T., Kimura, R. S., Paul, D. L., Takasaka, T., and Adams, J. C. (2000). Gap junction systems in the mammalian cochlea. *Brain Res.* **32**, 163–166.

King, L. S., Choi, M., Fernandez, P. C., Cartron, J. P., and Agre, P. (2001). Defective urinary-concentrating ability due to a complete deficiency of aquaporin-1. *N. Engl. J. Med.* **345**, 175–179.

Kiselyov, K., Chen, J., Rbaibi, Y., Oberdick, D., Tjon-Kon-Sang, S., Shcheynikov, N., Muallem, S., and Soyombo, A. (2005). TRP-ML1 is a lysosomal monovalent cation channel that undergoes proteolytic cleavage. *J. Biol. Chem.* **280**, 43218–43223.

Ko, S. B., Zeng, W., Dorwart, M. R., Luo, X., Kim, K. H., Millen, L., Goto, H., Naruse, S., Soyombo, A., Thomas, P. J., and Muallem, S. (2004). Gating of CFTR by the STAS domain of SLC26 transporters. *Nat. Cell Biol.* **6**, 343–350.

Kokubo, Y., Iwai, N., Tago, N., Inamoto, N., Okayama, A., Yamawaki, H., Naraba, H., and Tomoike, H. (2005). Association analysis between hypertension and CYBA, CLCNKB, and KCNMB1 functional polymorphisms in the Japanese population—The Suita Study. *Circ. J.* **69**, 138–142.

Konstan, M. W., Hilliard, K. A., Norvell, T. M., and Berger, M. (1994). Bronchoalveolar lavage findings in cystic fibrosis patients with stable, clinically mild lung disease suggest ongoing infection and inflammation. *Am. J. Respir. Crit. Care Med.* **150**, 448–454.

Kornak, U., Schulz, A., Friedrich, W., Uhlhaas, S., Kremens, B., Voit, T., Hasan, C., Bode, U., Jentsch, T. J., and Kubisch, C. (2000). Mutations in the a3 subunit of the vacuolar H⁺-ATPase cause infantile malignant osteopetrosis. *Hum. Mol. Genet.* **9**, 2059–2063.

Kornak, U., Kasper, D., Bösl, M. R., Kaiser, E., Schweizer, M., Schulz, A., Friedrich, W., Delling, G., and Jentsch, T. J. (2001). Loss of the ClC-7 chloride channel leads to osteopetrosis in mice and man. *Cell* **104**, 205–215.

Koulen, P., Cai, Y., Geng, L., Maeda, Y., Nishimura, S., Witzgall, R., Ehrlich, B. E., and Somlo, S. (2002). Polycystin-2 is an intracellular calcium release channel. *Nat. Cell Biol.* **4**, 191–197.

Kros, C. J., Ruppersberg, J. P., and Rusch, A. (1998). Expression of a potassium current in inner hair cells during development of hearing in mice. *Nature* **394**, 281–284.

Kubisch, C., Schroeder, B. C., Friedrich, T., Lütjohann, B., El-Amraoui, A., Marlin, S., Petit, C., and Jentsch, T. J. (1999). KCNQ4, a novel potassium channel expressed in sensory outer hair cells, is mutated in dominant deafness. *Cell* **96**, 437–446.

Lange, P. F., Wartosch, L., Jentsch, T. J., and Fuhrmann, J. C. (2006). ClC-7 requires Ostm1 as a beta-subunit to support bone resorption and lysosomal function. *Nature* **440**, 220–223.

LaPlante, J. M., Falardeau, J., Sun, M., Kanazirska, M., Brown, E. M., Slaugenhaupt, S. A., and Vassilev, P. M. (2002). Identification and characterization of the single channel function of human mucolipin-1 implicated in mucolipidosis type IV, a disorder affecting the lysosomal pathway. *FEBS Lett.* **532**, 183–187.

Liou, T. G., Adler, F. R., Cox, D. R., and Cahill, B. C. (2007). Lung transplantation and survival in children with cystic fibrosis. *N. Engl. J. Med.* **357**, 2143–2152.

Lloyd, S. E., Pearce, S. H., Fisher, S. E., Steinmeyer, K., Schwappach, B., Scheinman, S. J., Harding, B., Bolino, A., Devoto, M., Goodyer, P., *et al.* (1996). A common molecular basis for three inherited kidney stone diseases. *Nature* **379**, 445–449.

Low, S. H., Vasanth, S., Larson, C. H., Mukherjee, S., Sharma, N., Kinter, M. T., Kane, M. E., Obara, T., and Weimbs, T. (2006). Polycystin-1, STAT6, and P100 function in a pathway that transduces ciliary mechanosensation and is activated in polycystic kidney disease. *Dev. Cell* **10**, 57–69.

Mall, M., Grubb, B. R., Harkema, J. R., O'Neal, W. K., and Boucher, R. C. (2004). Increased airway epithelial Na^+ absorption produces cystic fibrosis-like lung disease in mice. *Nat. Med.* **10**, 487–493.

McDonald, F. J., Yang, B., Hrstka, R. F., Drummond, H. A., Tarr, D. E., McCray, P. B., Jr, Stokes, J. B., Welsh, M. J., and Williamson, R. A. (1999). Disruption of the beta subunit of the epithelial Na^+ channel in mice: Hyperkalemia and neonatal death associated with a pseudohypoaldosteronism phenotype. *Proc. Natl Acad. Sci. USA* **96**, 1727–1731.

Meij, I. C., Koenderink, J. B., van Bokhoven, H., Assink, K. F., Groenestege, W. T., de Pont, J. J., Bindels, R. J., Monnens, L. A., van den Heuvel, L. P., and Knoers, N. V. (2000). Dominant isolated renal magnesium loss is caused by misrouting of the $Na(+),K(+)$-ATPase gamma-subunit. *Nat. Genet.* **26**, 265–266.

Merin, S., Livni, N., Berman, E. R., and Yatziv, S. (1975). Mucolipidosis IV: Ocular, systemic, and ultrastructural findings. *Invest. Ophthalmol.* **14**, 437–448.

Mohammad-Panah, R., Gyomorey, K., Rommens, J., Choudhury, M., Li, C., Wang, Y., and Bear, C. E. (2001). ClC-2 contributes to native chloride secretion by a human intestinal cell line, Caco-2. *J. Biol. Chem.* **276**, 8306–8313.

Moller, C. C., Wei, C., Altintas, M. M., Li, J., Greka, A., Ohse, T., Pippin, J. W., Rastaldi, M. P., Wawersik, S., Schiavi, S., *et al.* (2007). Induction of TRPC6 channel in acquired forms of proteinuric kidney disease. *J. Am. Soc. Nephrol.* **18**, 29–36.

Murray, C. B., Morales, M. M., Flotte, T. R., McGrath-Morrow, S. A., Guggino, W. B., and Zeitlin, P. L. (1995). ClC-2: A developmentally dependent chloride channel expressed in the fetal lung and downregulated after birth. *Am. J. Respir. Cell Mol. Biol.* **12**, 597–604.

Myerburg, M. M., Butterworth, M. B., McKenna, E. E., Peters, K. W., Frizzell, R. A., Kleyman, T. R., and Pilewski, J. M. (2006). Airway surface liquid volume regulates ENaC by altering the serine protease–protease inhibitor balance: A mechanism for sodium hyperabsorption in cystic fibrosis. *J. Biol. Chem.* **281**, 27942–27949.

Nauli, S. M., Alenghat, F. J., Luo, Y., Williams, E., Vassilev, P., Li, X., Elia, A. E., Lu, W., Brown, E. M., Quinn, S. J., *et al.* (2003). Polycystins 1 and 2 mediate mechanosensation in the primary cilium of kidney cells. *Nat. Genet.* **33**, 129–137.

Neyroud, N., Tesson, F., Denjoy, I., Leibovici, M., Donger, C., Barhanin, J., Faure, S., Gary, F., Coumel, P., Petit, C., *et al.* (1997). A novel mutation in the potassium channel gene *KVLQT1* causes the Jervell and Lange–Nielsen cardioauditory syndrome. *Nat. Genet.* **15**, 186–189.

Nilius, B., Voets, T., and Peters, J. (2005). TRP channels in disease. *Sci. STKE* **2005**, re8.

Nozu, K., Inagaki, T., Fu, X. J., Nozu, Y., Kaito, H., Kanda, K., Sekine, T., Igarashi, T., Nakanishi, K., Yoshikawa, N., *et al.* (2008). Molecular analysis of digenic inheritance in Bartter syndrome with sensorineural deafness. *J. Med. Genet.* **45**, 182–186.

Pangrazio, A., Poliani, P. L., Megarbane, A., Lefranc, G., Lanino, E., Di Rocco, M., Rucci, F., Lucchini, F., Ravanini, M., Facchetti, F., et al. (2006). Mutations in OSTM1 (grey lethal) define a particularly severe form of autosomal recessive osteopetrosis with neural involvement. *J. Bone Miner. Res.* **21,** 1098–1105.

Pedemonte, N., Lukacs, G. L., Du, K., Caci, E., Zegarra-Moran, O., Galietta, L. J., and Verkman, A. S. (2005). Small-molecule correctors of defective DeltaF508-CFTR cellular processing identified by high-throughput screening. *J. Clin. Invest.* **115,** 2564–2571.

Pena-Munzenmayer, G., Catalan, M., Cornejo, I., Figueroa, C. D., Melvin, J. E., Niemeyer, M. I., Cid, L. P., and Sepulveda, F. V. (2005). Basolateral localization of native ClC-2 chloride channels in absorptive intestinal epithelial cells and basolateral sorting encoded by a CBS-2 domain di-leucine motif. *J. Cell Sci.* **118,** 4243–4252.

Pennekamp, P., Karcher, C., Fischer, A., Schweickert, A., Skryabin, B., Horst, J., Blum, M., and Dworniczak, B. (2002). The ion channel polycystin-2 is required for left-right axis determination in mice. *Curr. Biol.* **12,** 938–943.

Persu, A., Duyme, M., Pirson, Y., Lens, X. M., Messiaen, T., Breuning, M. H., Chauveau, D., Levy, M., Grunfeld, J. P., and Devuyst, O. (2004). Comparison between siblings and twins supports a role for modifier genes in ADPKD. *Kidney Int.* **66,** 2132–2136.

Peters, D. J., Spruit, L., Saris, J. J., Ravine, D., Sandkuijl, L. A., Fossdal, R., Boersma, J., van Eijk, R., Norby, S., Constantinou-Deltas, C. D., et al. (1993). Chromosome 4 localization of a second gene for autosomal dominant polycystic kidney disease. *Nat. Genet.* **5,** 359–362.

Picollo, A., and Pusch, M. (2005). Chloride/proton antiporter activity of mammalian CLC proteins ClC-4 and ClC-5. *Nature* **436,** 420–423.

Piwon, N., Günther, W., Schwake, M., Bösl, M. R., and Jentsch, T. J. (2000). ClC-5 Cl⁻-channel disruption impairs endocytosis in a mouse model for Dent's disease. *Nature* **408,** 369–373.

Poët, M., Kornak, U., Schweizer, M., Zdebik, A. A., Scheel, O., Hoelter, S., Wurst, W., Schmitt, A., Fuhrmann, J. C., Planells-Cases, R., et al. (2006). Lysosomal storage disease upon disruption of the neuronal chloride transport protein ClC-6. *Proc. Natl Acad. Sci. USA* **103,** 13854–13859.

Popli, K., and Stewart, J. (2007). Infertility and its management in men with cystic fibrosis: Review of literature and clinical practices in the UK. *Hum. Fertil. (Camb., Engl.)* **10,** 217–221.

Preston, G. M., and Agre, P. (1991). Isolation of the cDNA for erythrocyte integral membrane protein of 28 kilodaltons: Member of an ancient channel family. *Proc. Natl Acad. Sci. USA* **88,** 11110–11114.

Preston, G. M., Carroll, T. P., Guggino, W. B., and Agre, P. (1992). Appearance of water channels in Xenopus oocytes expressing red cell CHIP28 protein. *Science* **256,** 385–387.

Qian, F., Germino, F. J., Cai, Y., Zhang, X., Somlo, S., and Germino, G. G. (1997). PKD1 interacts with PKD2 through a probable coiled-coil domain. *Nat. Genet.* **16,** 179–183.

Qian, Q., Hunter, L. W., Li, M., Marin-Padilla, M., Prakash, Y. S., Somlo, S., Harris, P. C., Torres, V. E., and Sieck, G. C. (2003). Pkd2 haploinsufficiency alters intracellular calcium regulation in vascular smooth muscle cells. *Hum. Mol. Genet.* **12,** 1875–1880.

Ramirez, A., Faupel, J., Goebel, I., Stiller, A., Beyer, S., Stockle, C., Hasan, C., Bode, U., Kornak, U., and Kubisch, C. (2004). Identification of a novel mutation in the coding region of the grey-lethal gene OSTM1 in human malignant infantile osteopetrosis. *Hum. Mutat.* **23,** 471–476.

Raychowdhury, M. K., Gonzalez-Perrett, S., Montalbetti, N., Timpanaro, G. A., Chasan, B., Goldmann, W. H., Stahl, S., Cooney, A., Goldin, E., and Cantiello, H. F. (2004). Molecular pathophysiology of mucolipidosis type IV: pH dysregulation of the mucolipin-1 cation channel. *Hum. Mol. Genet.* **13,** 617–627.

Reddy, M. M., Light, M. J., and Quinton, P. M. (1999). Activation of the epithelial Na⁺ channel (ENaC) requires CFTR Cl⁻ channel function. *Nature* **402,** 301–304.

Reeders, S. T., Breuning, M. H., Davies, K. E., Nicholls, R. D., Jarman, A. P., Higgs, D. R., Pearson, P. L., and Weatherall, D. J. (1985). A highly polymorphic DNA marker linked to adult polycystic kidney disease on chromosome 16. *Nature* **317,** 542–544.

Reidy, K., and Kaskel, F. J. (2007). Pathophysiology of focal segmental glomerulosclerosis. *Pediatr. Nephrol. (Berl., Germany)* **22,** 350–354.

Reinalter, S. C., Jeck, N., Brochhausen, C., Watzer, B., Nusing, R. M., Seyberth, H. W., and Komhoff, M. (2002). Role of cyclooxygenase-2 in hyperprostaglandin E syndrome/antenatal Bartter syndrome. *Kidney Int.* **62,** 253–260.

Reiser, J., Polu, K. R., Moller, C. C., Kenlan, P., Altintas, M. M., Wei, C., Faul, C., Herbert, S., Villegas, I., Avila-Casado, C., *et al.* (2005). TRPC6 is a glomerular slit diaphragm-associated channel required for normal renal function. *Nat. Genet.* **37,** 739–744.

Riordan, J. R., Rommens, J. M., Kerem, B., Alon, N., Rozmahel, R., Grzelczak, Z., Zielenski, J., Lok, S., Plavsic, N., Chou, J. L., *et al.* (1989). Identification of the cystic fibrosis gene: Cloning and characterization of complementary DNA. *Science* **245,** 1066–1073.

Rosenstein, B. J., and Cutting, G. R. (1998). The diagnosis of cystic fibrosis: A consensus statement. Cystic Fibrosis Foundation Consensus Panel. *J. Pediatr.* **132,** 589–595.

Rosenthal, W., Seibold, A., Antaramian, A., Lonergan, M., Arthus, M. F., Hendy, G. N., Birnbaumer, M., and Bichet, D. G. (1992). Molecular identification of the gene responsible for congenital nephrogenic diabetes insipidus. *Nature* **359,** 233–235.

Rossier, B. C., Pradervand, S., Schild, L., and Hummler, E. (2002). Epithelial sodium channel and the control of sodium balance: Interaction between genetic and environmental factors. *Annu. Rev. Physiol.* **64,** 877–897.

Rufo, P. A., Jiang, L., Moe, S. J., Brugnara, C., Alper, S. L., and Lencer, W. I. (1996). The antifungal antibiotic, clotrimazole, inhibits Cl^- secretion by polarized monolayers of human colonic epithelial cells. *J. Clin. Invest.* **98,** 2066–2075.

Ruttiger, L., Sausbier, M., Zimmermann, U., Winter, H., Braig, C., Engel, J., Knirsch, M., Arntz, C., Langer, P., Hirt, B., *et al.* (2004). Deletion of the Ca^{2+}-activated potassium (BK) alpha-subunit but not the BKbeta1-subunit leads to progressive hearing loss. *Proc. Natl Acad. Sci. USA* **101,** 12922–12927.

Scheel, O., Zdebik, A. A., Lourdel, S., and Jentsch, T. J. (2005). Voltage-dependent electrogenic chloride/proton exchange by endosomal CLC proteins. *Nature* **436,** 424–427.

Schild, L., Schneeberger, E., Gautschi, I., and Firsov, D. (1997). Identification of amino acid residues in the alpha, beta, and gamma subunits of the epithelial sodium channel (ENaC) involved in amiloride block and ion permeation. *J. Gen. Physiol.* **109,** 15–26.

Schlingmann, K. P., Weber, S., Peters, M., Niemann Nejsum, L., Vitzthum, H., Klingel, K., Kratz, M., Haddad, E., Ristoff, E., Dinour, D., *et al.* (2002). Hypomagnesemia with secondary hypocalcaemia is caused by mutations in TRPM6, a new member of the TRPM gene family. *Nat. Genet.* **31,** 166–170.

Schlingmann, K. P., Konrad, M., Jeck, N., Waldegger, P., Reinalter, S. C., Holder, M., Seyberth, H. W., and Waldegger, S. (2004). Salt wasting and deafness resulting from mutations in two chloride channels. *N. Engl. J. Med.* **350,** 1314–1319.

Schnermann, J., Chou, C. L., Ma, T., Traynor, T., Knepper, M. A., and Verkman, A. S. (1998). Defective proximal tubular fluid reabsorption in transgenic aquaporin-1 null mice. *Proc. Natl Acad. Sci. USA* **95,** 9660–9664.

Schroeder, B. C., Waldegger, S., Fehr, S., Bleich, M., Warth, R., Greger, R., and Jentsch, T. J. (2000). A constitutively open potassium channel formed by KCNQ1 and KCNE3. *Nature* **403,** 196–199.

Schulze-Bahr, E., Wang, Q., Wedekind, H., Haverkamp, W., Chen, Q., Sun, Y., Rubie, C., Hordt, M., Towbin, J. A., Borggrefe, M., *et al.* (1997). *KCNE1* mutations cause Jervell and Lange–Nielsen syndrome. *Nat. Genet.* **17,** 267–268.

Schwake, M., Friedrich, T., and Jentsch, T. J. (2001). An internalization signal in ClC-5, an endosomal Cl-channel mutated in dent's disease. *J. Biol. Chem.* **276,** 12049–12054.

Serohijos, A. W., Hegedus, T., Aleksandrov, A. A., He, L., Cui, L., Dokholyan, N. V., and Riordan, J. R. (2008). Phenylalanine-508 mediates a cytoplasmic-membrane domain contact in the CFTR 3D structure crucial to assembly and channel function. *Proc. Natl Acad. Sci. USA* **105,** 3256–3261.

Sheridan, M. B., Fong, P., Groman, J. D., Conrad, C., Flume, P., Diaz, R., Harris, C., Knowles, M., and Cutting, G. R. (2005). Mutations in the beta-subunit of the epithelial Na^+ channel in patients with a cystic fibrosis-like syndrome. *Hum. Mol. Genet.* **14,** 3493–3498.

Shiels, A., and Bassnett, S. (1996). Mutations in the founder of the MIP gene family underlie cataract development in the mouse. *Nat. Genet.* **12,** 212–215.

Simon, D. B., Karet, F. E., Hamdan, J. M., DiPietro, A., Sanjad, S. A., and Lifton, R. P. (1996a). Bartter's syndrome, hypokalaemic alkalosis with hypercalciuria, is caused by mutations in the Na–K–2Cl cotransporter NKCC2. *Nat. Genet.* **13,** 183–188.

Simon, D. B., Karet, F. E., Rodriguez-Soriano, J., Hamdan, J. H., DiPietro, A., Trachtman, H., Sanjad, S. A., and Lifton, R. P. (1996b). Genetic heterogeneity of Bartter's syndrome revealed by mutations in the K^+ channel, ROMK. *Nat. Genet.* **14,** 152–156.

Simon, D. B., Nelson-Williams, C., Bia, M. J., Ellison, D., Karet, F. E., Molina, A. M., Vaara, I., Iwata, F., Cushner, H. M., Koolen, M., *et al.* (1996c). Gitelman's variant of Bartter's syndrome, inherited hypokalaemic alkalosis, is caused by mutations in the thiazide-sensitive Na–Cl cotransporter. *Nat. Genet.* **12,** 24–30.

Simon, D. B., Bindra, R. S., Mansfield, T. A., Nelson-Williams, C., Mendonca, E., Stone, R., Schurman, S., Nayir, A., Alpay, H., Bakkaloglu, A., *et al.* (1997). Mutations in the chloride channel gene, CLCNKB, cause Bartter's syndrome type III. *Nat. Genet.* **17,** 171–178.

Simon, D. B., Lu, Y., Choate, K. A., Velazquez, H., Al-Sabban, E., Praga, M., Casari, G., Bettinelli, A., Colussi, G., Rodriguez-Soriano, J., *et al.* (1999). Paracellin-1, a renal tight junction protein required for paracellular Mg^{2+} resorption. *Science* **285,** 103–106.

Smith, B. L., and Agre, P. (1991). Erythrocyte Mr 28,000 transmembrane protein exists as a multi-subunit oligomer similar to channel proteins. *J. Biol. Chem.* **266,** 6407–6415.

Sohara, E., Rai, T., Yang, S. S., Uchida, K., Nitta, K., Horita, S., Ohno, M., Harada, A., Sasaki, S., and Uchida, S. (2006). Pathogenesis and treatment of autosomal-dominant nephrogenic diabetes insipidus caused by an aquaporin 2 mutation. *Proc. Natl Acad. Sci. USA* **103,** 14217–14222.

Soyombo, A. A., Tjon-Kon-Sang, S., Rbaibi, Y., Bashllari, E., Bisceglia, J., Muallem, S., and Kiselyov, K. (2006). TRP-ML1 regulates lysosomal pH and acidic lysosomal lipid hydrolytic activity. *J. Biol. Chem.* **281,** 7294–7301.

Speirs, H. J., Wang, W. Y., Benjafield, A. V., and Morris, B. J. (2005). No association with hypertension of CLCNKB and TNFRSF1B polymorphisms at a hypertension locus on chromosome 1p36. *J. Hypertens.* **23,** 1491–1496.

Staub, O., Dho, S., Henry, P., Correa, J., Ishikawa, T., McGlade, J., and Rotin, D. (1996). WW domains of Nedd4 bind to the proline-rich PY motifs in the epithelial Na^+ channel deleted in Liddle's syndrome. *EMBO J.* **15,** 2371–2380.

Staub, O., Gautschi, I., Ishikawa, T., Breitschopf, K., Ciechanover, A., Schild, L., and Rotin, D. (1997). Regulation of stability and function of the epithelial Na^+ channel (ENaC) by ubiquitination. *EMBO J.* **16,** 6325–6336.

Steinmeyer, K., Schwappach, B., Bens, M., Vandewalle, A., and Jentsch, T. J. (1995). Cloning and functional expression of rat CLC-5, a chloride channel related to kidney disease. *J. Biol. Chem.* **270,** 31172–31177.

Stobrawa, S. M., Breiderhoff, T., Takamori, S., Engel, D., Schweizer, M., Zdebik, A. A., Bösl, M. R., Ruether, K., Jahn, H., Draguhn, A., *et al.* (2001). Disruption of ClC-3, a chloride channel expressed on synaptic vesicles, leads to a loss of the hippocampus. *Neuron* **29,** 185–196.

Strausbaugh, S. D., and Davis, P. B. (2007). Cystic fibrosis: A review of epidemiology and pathobiology. *Clin. Chest Med.* **28,** 279–288.

Stutts, M. J., Canessa, C. M., Olsen, J. C., Hamrick, M., Cohn, J. A., Rossier, B. C., and Boucher, R. C. (1995). CFTR as a cAMP-dependent regulator of sodium channels. *Science* **269,** 847–850.

Sun, M., Goldin, E., Stahl, S., Falardeau, J. L., Kennedy, J. C., Acierno, J. S., Jr., Bove, C., Kaneski, C. R., Nagle, J., Bromley, M. C., *et al.* (2000). Mucolipidosis type IV is caused by mutations in a gene encoding a novel transient receptor potential channel. *Hum. Mol. Genet.* **9,** 2471–2478.

Suzuki, T., Rai, T., Hayama, A., Sohara, E., Suda, S., Itoh, T., Sasaki, S., and Uchida, S. (2006). Intracellular localization of ClC chloride channels and their ability to form hetero-oligomers. *J. Cell. Physiol.* **206**, 792–798.

Taitelbaum-Swead, R., Brownstein, Z., Muchnik, C., Kishon-Rabin, L., Kronenberg, J., Megirov, L., Frydman, M., Hildesheimer, M., and Avraham, K. B. (2006). Connexin-associated deafness and speech perception outcome of cochlear implantation. *Arch. Otolaryngol. Head Neck Surg.* **132**, 495–500.

Teubner, B., Michel, V., Pesch, J., Lautermann, J., Cohen-Salmon, M., Sohl, G., Jahnke, K., Winterhager, E., Herberhold, C., Hardelin, J. P., *et al.* (2003). Connexin30 (Gjb6)-deficiency causes severe hearing impairment and lack of endocochlear potential. *Hum. Mol. Genet.* **12**, 13–21.

Treusch, S., Knuth, S., Slaugenhaupt, S. A., Goldin, E., Grant, B. D., and Fares, H. (2004). *Caenorhabditis elegans* functional orthologue of human protein h-mucolipin-1 is required for lysosome biogenesis. *Proc. Natl Acad. Sci. USA* **101**, 4483–4488.

Tyson, J., Tranebjaerg, L., Bellman, S., Wren, C., Taylor, J. F., Bathen, J., Aslaksen, B., Sorland, S. J., Lund, O., Malcolm, S., *et al.* (1997). IsK and K$_v$LQT1: Mutation in either of the two subunits of the slow component of the delayed rectifier potassium channel can cause Jervell and Lange–Nielsen syndrome. *Hum. Mol. Genet.* **6**, 2179–2185.

Venugopal, B., Browning, M. F., Curcio-Morelli, C., Varro, A., Michaud, N., Nanthakumar, N., Walkley, S. U., Pickel, J., and Slaugenhaupt, S. A. (2007). Neurologic, gastric, and opthalmologic pathologies in a murine model of mucolipidosis type IV. *Am. J. Hum. Genet.* **81**, 1070–1083.

Vetter, D. E., Mann, J. R., Wangemann, P., Liu, J., McLaughlin, K. J., Lesage, F., Marcus, D. C., Lazdunski, M., Heinemann, S. F., and Barhanin, J. (1996). Inner ear defects induced by null mutation of the isk gene. *Neuron* **17**, 1251–1264.

Waguespack, S. G., Koller, D. L., White, K. E., Fishburn, T., Carn, G., Buckwalter, K. A., Johnson, M., Kocisko, M., Evans, W. E., Foroud, T., and Econs, M. J. (2003). Chloride channel 7 (ClCN7) gene mutations and autosomal dominant osteopetrosis, type II. *J. Bone Miner. Res.* **18**, 1513–1518.

Walder, R. Y., Landau, D., Meyer, P., Shalev, H., Tsolia, M., Borochowitz, Z., Boettger, M. B., Beck, G. E., Englehardt, R. K., Carmi, R., and Sheffield, V. C. (2002). Mutation of TRPM6 causes familial hypomagnesemia with secondary hypocalcaemia. *Nat. Genet.* **31**, 171–174.

Wang, S. S., Devuyst, O., Courtoy, P. J., Wang, X. T., Wang, H., Wang, Y., Thakker, R. V., Guggino, S., and Guggino, W. B. (2000). Mice lacking renal chloride channel, CLC-5, are a model for Dent's disease, a nephrolithiasis disorder associated with defective receptor-mediated endocytosis. *Hum. Mol. Genet.* **9**, 2937–2945.

Watnick, T. J., Torres, V. E., Gandolph, M. A., Qian, F., Onuchic, L. F., Klinger, K. W., Landes, G., and Germino, G. G. (1998). Somatic mutation in individual liver cysts supports a two-hit model of cystogenesis in autosomal dominant polycystic kidney disease. *Mol. Cell* **2**, 247–251.

Weissmann, N., Dietrich, A., Fuchs, B., Kalwa, H., Ay, M., Dumitrascu, R., Olschewski, A., Storch, U., Mederos y Schnitzler, M., Ghofrani, H. A., *et al.* (2006). Classical transient receptor potential channel 6 (TRPC6) is essential for hypoxic pulmonary vasoconstriction and alveolar gas exchange. *Proc. Natl Acad. Sci. USA* **103**, 19093–19098.

Wilschanski, M., Yahav, Y., Yaacov, Y., Blau, H., Bentur, L., Rivlin, J., Aviram, M., Bdolah-Abram, T., Bebok, Z., Shushi, L., *et al.* (2003). Gentamicin-induced correction of CFTR function in patients with cystic fibrosis and CFTR stop mutations. *N. Engl. J. Med.* **349**, 1433–1441.

Winn, M. P., Conlon, P. J., Lynn, K. L., Farrington, M. K., Creazzo, T., Hawkins, A. F., Daskalakis, N., Kwan, S. Y., Ebersviller, S., Burchette, J. L., *et al.* (2005). A mutation in the TRPC6 cation channel causes familial focal segmental glomerulosclerosis. *Science* **308**, 1801–1804.

Wrong, O. M., Norden, A. G., and Feest, T. G. (1994). Dent's disease; a familial proximal renal tubular syndrome with low-molecular-weight proteinuria, hypercalciuria, nephrocalcinosis, metabolic bone disease, progressive renal failure and a marked male predominance. *Q. J. Med.* **87**, 473–493.

Yoshikawa, M., Uchida, S., Ezaki, J., Rai, T., Hayama, A., Kobayashi, K., Kida, Y., Noda, M., Koike, M., Uchiyama, Y., *et al.* (2002). CLC-3 deficiency leads to phenotypes similar to human neuronal ceroid lipofuscinosis. *Genes Cells* **7,** 597–605.

Yu, Y., Fantozzi, I., Remillard, C. V., Landsberg, J. W., Kunichika, N., Platoshyn, O., Tigno, D. D., Thistlethwaite, P. A., Rubin, L. J., and Yuan, J. X. (2004). Enhanced expression of transient receptor potential channels in idiopathic pulmonary arterial hypertension. *Proc. Natl Acad. Sci. USA* **101,** 13861–13866.

Zdebik, A. A., Cuffe, J. E., Bertog, M., Korbmacher, C., and Jentsch, T. J. (2004). Additional disruption of the ClC-2 Cl(−) channel does not exacerbate the cystic fibrosis phenotype of cystic fibrosis transmembrane conductance regulator mouse models. *J. Biol. Chem.* **279,** 22276–22283.

Index

A

ADPKD. *See* Autosomal-dominant polycystic kidney disease
ATP-binding cassette (ABC) transporters, 117
Autosomal-dominant myotonia congenital, 27–28
Autosomal-dominant polycystic kidney disease, 123–124
Autosomal-recessive neurodegenerative lysosomal storage disorder, 137–138

B

Bartter syndrome, 124–126
Becker myotonia, 28–29
Bone sclerosis, 142

C

Channelopathies affecting renal function
autosomal-dominant polycystic kidney disease (ADPKD), 123–124
focal segmental glomerulosclerosis (FSGS)
familial forms of, 122
filtration apparatus, 121
tubular disorders

Bartter syndrome, 124–126
familial hypomagnesemia, 129–130
nephrogenic diabetes insipidus, 128–129
renal NaCl balance to ENaC, 126–128
X-linked hypercalciuric nephrolithiasis, 130–132
Congenital insensitivity, inherited pain disorders
historical background and symptom, 101
linkage analysis, 101–102
molecular genetics, 102
molecular pathophysiology, 102–103
Connexin genes, 136–137
Cortical spreading depression (CSD)
in experimental animals, 61–63
migraine aura, 62
Cystic fibrosis transmembrane regulator (CFTR)
ATP-binding cassette (ABC) transporters, 117
definition, 116
diagnosis of, 118
immunocytochemistry, 120
molecular genetics, 116–117
pancreatic gland, 119
role in epithelial transport, 117
treatment and prevention of pulmonary complications, 119–120

D

Dent's disease. *See* X-linked hypercalciuric nephrolithiasis

F

Familial hemiplegic migraine
  characterization, 65–66
  FHM1 *CACNA1A* gene, 66
  FHM2 *ATP1A2* gene, 66–67
  FHM3 *SCN1A* gene, 67
  FHM4 and beyond, 68
  ionopathy
    FHM1, 73–74
    FHM2, 74–75
    FHM3, 75
  mutations
    CaV2.1 (P/Q-type) calcium channel (FHM1), 68–71
    Na+/K+-ATPase α2 transporter (FHM2), 71–72
    NaV1.1 (FHM3), 72–73
  sporadic hemiplegic migraine, 67–68
Familial hyperPP, 11
Familial hypomagnesemia, 129–130
Familial hypoPP, 5
Focal segmental glomerulosclerosis (FSGS)
  familial forms of, 122
  filtration apparatus, 121

G

Gene therapy, myotonia congenita, 49
Genetic disorder and migraine
  environmental influences in migraine susceptibility, 64
  gene discovery, 64–65
  genetic epidemiology, 63–64

H

Hearing loss and channelopathies
  auditory deprivation, 132
  connexin genes, 136–137
  hair cells, 135–136
  K^+ recycling model of inner ear, 133–134
  prelingual deafness, 132–133
  sensory hair cells, 134–135
Hyperkalemic periodic paralysis (HyperPP)
  definition, 11
  muscle paralysis, 13–14
  voltage-gated sodium channels, 13
  vs. hypokalemic periodic paralyses, 12–13
Hypokalemic periodic paralyses (HypoPP)
  definition, 4
  familial type, 5–6
  genetic linkage studies, 9
  muscle weakness, 6
  thyrotoxic type, 5
  treatment, 17–18
  vacuoles, 7
  voltage-gated calcium channel, 9
  voltage-gated sodium channel, 9–11

I

Inherited erythromelalgia (IEM)
  linkage analysis, 93–94
  molecular genetics, 94–95
  mutation-induced changes
    biophysical properties, 95–96
    DRG neuron firing, 96–98
  types, symptoms and management, 92–93
Inherited pain disorders
  congenital insensitivity
    historical background and symptom, 101

linkage analysis, 101–102
molecular genetics, 102
molecular pathophysiology, 102–103
inherited erythromelalgia
biophysical properties, 95–96
DRG neuron firing, 96–98
linkage analysis, 93–94
molecular genetics, 94–95
types, symptoms and
management, 92–93
paroxysmal extreme pain disorder
(PEPD), 99–101
Ion channels
diseases, 114, 116
functional characteristics of, 114
mutations, 15–16
Ionopathy
FHM1, 73–74
FHM2, 74–75
FHM3, 75

M

Migraine
clinical phases and pathophysiology, 61
comorbidity, 60–61
cortical spreading depression (CSD)
in experimental animals, 61–63
migraine aura, 62
disabling episodic disorder, 58–59
epidemiology of, 59–60
genetic disorder
environmental influences in migraine
susceptibility, 64
gene discovery, 64–65
genetic epidemiology, 63–64
Molecular genetic testing, 32
Molecular genetics
inherited pain disorders
congenital insensitivity, 102

inherited erythromelalgia
(IEM), 94–95
myotonia congenita
muscle chloride channel ClC-1, 32–33
spectrum of CLCN1 mutations, 33–41
Mucolipidosis type IV
definition, 137
in *Caenorhabditis elegans*, 138
MCOLN1 codes, 137–138
Myotonia congenita
clinical aspects of
autosomal-dominant myotonia
congenital, 27–28
diagnosis of, 29–32
recessive generalized myotonia, 28–29
in animal models, 46–47
molecular genetics of
muscle chloride channel ClC-1, 32–33
spectrum of CLCN1 mutations, 33–41
physiological basis of
CLCN1 mutations, 43–45
sarcolemmal chloride
conductance, 42–43
treatment of
gene therapy for, 49
nonpharmacological therapies, 48–49
pharmacological therapies, 47–48

N

Na$_v$ 1.7, sodium channel
discovery and electrophysiological
properties, 89–90
in pain syndrome, animal studies
inflammation-induced pain, 91
streptozotocin-induced diabetic
neuropathy (STZ), 91
traumatic nerve injury, 90–91
inherited pain disorders
congenital insensitivity, 101–103

Na$_v$ 1.7, sodium channel (*cont.*)
  inherited erythromelalgia, 92–98
  paroxysmal extreme pain disorder
    (PEPD), 99–101
Nephrogenic diabetes insipidus, 128–129
Neuronal ceroid lipofuscinosis (NCL), 140

O

Osteopetrosis and lysosomal storage disease
  bone sclerosis, 142
  ClC-7, 139
  CLCN7 gene, 138–139
  lysosomal pathologies, 140–141
  OSTM1 encodes, 141

P

Pain syndrome, Na$_v$ 1.7, animal studies
  inflammation-induced pain, 91
  streptozotocin-induced diabetic
    neuropathy (STZ), 91
  traumatic nerve injury, 90–91
Paroxysmal extreme pain disorder
  (PEPD), 99–101
Percussion myotonia, 30
Periodic paralysis
  classification, 4
  hyperkalemic periodic paralysis
    definition, 11
    muscle paralysis, 13–14
    voltage-gated sodium channels, 13
    vs. hypokalemic periodic
      paralyses, 12–13
  hypokalemic periodic paralysis
    definition, 4
    familial type, 5–6
    genetic linkage studies, 9
    muscle weakness, 6
    thyrotoxic type, 5
    treatment, 17–18

  vacuoles, 7
  voltage-gated calcium channel, 9
  voltage-gated sodium channel, 9–11
  ion channel mutations, 15–16
  KCNE3 gene, 14–15
  muscle weakness, 16
  normoPP, 14
  pathophysiology of, 12
Phenylalanine-508, 118

R

Recessive generalized myotonia
  (RGM), 28–29

S

SCN9A gene
  congenital insensitivity, 101–102
  for sodium channels, 88
  inherited erythromelalgia, 93–94
  paroxysmal extreme pain disorder, 99
Sporadic hemiplegic migraine,
  (SHM), 67–68
Streptozotocin-induced diabetic
  neuropathy (STZ), 91

T

Thomsen disease. *See* Autosomal-
  dominant myotonia congenita
Thyrotoxic hypoPP, 5
Tubular aggregates, 7
Tubular disorders
  Bartter syndrome, 124–126
  familial hypomagnesemia, 129–130
  nephrogenic diabetes insipidus, 128–129
  renal NaCl balance to ENaC, 126–128
  X-linked hypercalciuric
    nephrolithiasis, 130–132

V

Vacuoles, 7
Voltage-gated sodium channels
  subcellular distribution, 89
  tissue distribution, 87–88

X

X-linked hypercalciuric nephrolithiasis
  CLCN5 codes, 130–131
  parathormone (PTH), 132